High Thermal Conductivity Materials

Subhash L. Shindé
Jitendra S. Goela
Editors

High Thermal Conductivity Materials

With 133 Illustrations

 Springer

Editors

Subhash L. Shindé (Ed.)
Sandia National Labs.
P.O. Box 5800, MS 0603
Albuquerque, NM 87185
USA
slshind@sandia.gov

Jitendra S. Goela (Ed.)
Rohm and Haas Advanced Materials
185 New Boston Street
Woburn, MA 01801
USA
jgoela@rohmhaas.com

Library of Congress Cataloging-in-Publication Data
Shindé, Subhash L.
  High thermal conductivity materials / Subhash L. Shindé, Jitendra S. Goela.
    p. cm.
  Includes bibliographical references and index.
  ISBN 0-387-22021-6 (alk. paper)
  1. Heat--Conduction. 2. Materials--Thermal properties. I. Shindé, Subhash L. Goela,
Jitendra S. II. Title.

QC321.S44 2004
536'.2012—dc22   2004049159

ISBN-10: 0-387-22021-6   e-ISBN-: 0-387-25100-6

ISBN-13: 978-0387-22021-5

Printed on acid-free paper.

Printed in the United States of America.   (SBA/Techset)

9 8 7 6 5 4 3 2 1

springeronline.com

In loving Memory of
Dr. Bhagwat Sahai Verma (1920–2005),
Retired Dean of Engineering,
Banaras Hindu University,
Varanasi, India

Preface

The demand for efficient thermal management has increased substantially over the last decade in every imaginable area, be it a formula 1 racing car suddenly braking to decelerate from 200 to 50 mph going around a sharp corner, a space shuttle entering the earth's atmosphere, or an advanced microprocessor operating at a very high speed. The temperatures at the hot junctions are extremely high and the thermal flux can reach values higher than a few hundred to a thousand watts/cm^2 in these applications. To take a specific example of the microelectronics area, the chip heat flux for CMOS microprocessors, though moderate compared to the numbers mentioned above have already reached values close to 100 W/cm^2, and are projected to increase above 200 W/cm^2 over the next few years. Although the thermal management strategies for microprocessors do involve power optimization through improved design, it is extremely difficult to eliminate "hot spots" completely. This is where high thermal conductivity materials find most of their applications, as "heat spreaders". The high thermal conductivity of these materials allows the heat to be carried away from the "hot spots" very quickly in all directions thereby "spreading" the heat. Heat spreading reduces the heat flux density, and thus makes it possible to cool systems using standard cooling solutions like finned heat sinks with forced air cooling. A quick review of the available information indicates that the microprocessors heat fluxes are quickly reaching the 100 W/cm^2 values, which makes it very difficult to use conventional air cooling (see for example, "Thermal challenges in microprocessor testing", by P. Tadayan et al. Intel Technology Journal, Q3, 2000, and Chu, R., and Joshi, Y., Eds. "NEMI Technology Roadmap, National Electronics Manufacturing Initiative", Herndon, VA, 2002).

One approach to address this problem is to design and develop materials with higher thermal conductivities. This is possible by developing a detailed understanding of the thermal conduction mechanisms in these materials and studying how the processing and resulting microstructures affect their thermal properties. These aspects are the subject matter of review in this book.

We have chosen to review our current understanding of the conduction mechanisms in the high thermal conductivity materials, various techniques to measure the thermal conductivity accurately, and the processing and thermal conduction properties of a few candidate high thermal conductivity materials. This is by no means an exhaustive review, but the chapters authored by internationally known experts should provide a good review of the status of their field and form a sound basis for further studies in these areas.

The eight chapters in this book are arranged to provide a coherent theme starting from theory to understanding of practical materials, so a scientist would be able to optimize properties of these materials using basic concepts. In Chapter 1, Srivastava covers the thermal conduction mechanisms in non-metallic solids in some detail. The thermal conductivity expression derived is used to provide guidelines for choosing high thermal conductivity materials. Thermal conductivity results for various materials including diamond, carbon nanotubes, and various other forms of carbon are presented. The results are also extended to polycrystalline, and low dimensional systems. In Chapter 2, Morelli deals with the thermal conductivity of materials near their Debye temperatures. It also compares the results of a simple model to experimental data from various classes of crystal structures. Ashegi et al. discusses accurate characterization of thermal conductivity of various materials in Chapter 3. They review various thermal conductivity measurement techniques available to a researcher in detail, and also recommend techniques particularly suitable for high thermal conductivity materials like AlN, SiC, and diamond. In Chapter 4. Fournier reviews an elegant technique, perfected by her group, for measuring thermal conductivity on a very small spatial scale in heterogeneous materials. It is believed that this technique would be very important when evaluating thermal performance of complex systems. Virkar et al. provides the current status of the understanding of processing, and resultant thermal conductivity of aluminum nitride ceramics in Chapter 5. This chapter lays out the thermodynamic foundation for processing that will result in oxygen impurity removal from AlN, and thus increase its thermal conductivity. We hope that general application of these concepts will help researchers optimize thermal conductivity of a host of material systems. In Chapters 6 and 7, Goela et al. describe the details of CVD-SiC, and diamond materials processing and their properties. Here again the inter-relationship between the microstructure development through processing, and its effect on thermal conductivity is presented. Finally, in Chapter 8, Kwon et al. describe theoretical and experimental aspects of the thermal transport properties of carbon nanotubes. The strong carbon atom network in these novel materials lead not only to very unusual mechanical and electrical properties, but also to high thermal conductivity along the tube axis. We hope that the concepts described in these chapters will survive the test of time, and launch many curious scientists into their own forays in this field of highly interesting materials and their properties.

The editors would like to thank all the authors for their time and effort and the Springer, New York staff for their highly professional handling of the production of this volume. Subhash L. Shindé would like to acknowledge his colleagues for their critical review of these manuscripts.

Subhash L. Shindé
Jitendra S. Goela
July, 2005

Contents

Contributors

Editors:
Subhash L. Shindé
Sandia National Labs
P.O. Box 5800, MS 0603
Albuquerque, NM 87185
USA
slshind@sandia.gov

Jitendra S. Goela
Rohm and Haas Advanced Materials
185 New Boston Street
Woburn, MA 01801, USA
jgoela@rohmhaas.com

Chapter 1
Lattice Thermal Conduction
Mechanism in Solids

Gyaneshwar P. Srivastava
School of Physics
University of Exeter
Stocker Road, Exeter EX4 4QL,
United Kingdom
gps@excc.ex.ac.uk

Chapter 2
High Lattice Thermal Conductivity
Solids

Donald T. Morelli
Materials Group, Delphi Corporation
Research Labs.
Shelby Township, MI, 48315, USA
donald.t.morelli@delphi.com

Glen A. Slack
Rensselaer Polytechnic Institute (RPI)
110 8th St., Troy, NY 12180, USA

Chapter 3
Thermal Characterization of the
High-Thermal-Conductivity Dielectrics

Yizhang Yang
Mechanical Engineering Department
Carnegie Mellon University
5000 Forbes Avenue
Pittsburgh, PA 15213, USA
yizhangy@andrew.cmu.edu

Sadegh M. Sadeghipour
Mechanical Engineering Department
Carnegie Mellon University
5000 Forbes Avenue
Pittsburgh, PA 15213, USA
sms@andrew.cmu.edu

Wenjun Liu
Mechanical Engineering Department
Carnegie Mellon University
5000 Forbes Avenue
Pittsburgh, PA 15213, USA
wenjunl@andrew.cmu.edu

Mehdi Asheghi
Mechanical Engineering Department
Carnegie Mellon University
5000 Forbes Avenue
Pittsburgh, PA 15213, USA
masheghi@andrew.cmu.edu

Maxat Touzelbaev
Advanced Micro Devices
3625, Peterson Way, MS 58
Santa Clara, CA 95054, USA
maxat.touzelbaev@amd.com

Chapter 4
Thermal Wave Probing of High-
Conductivity Heterogeneous Materials

Danièle Fournier
Laboratoire UPR A005 du CNRS
ESPCI, 10 rue Vauquelin 75005 Paris,
France
fournier@optique.espci.fr

Chapter 5
Fabrication of High-Thermal-
Conductivity Polycrystalline Aluminum
Nitride: Thermodynamic and Kinetic
Aspects of Oxygen Removal

Anil V. Virkar
Materials Science and Engineering
Energy & Mineral Research Office
122 S Central Campus Drive,
Room 304
University of Utah
Salt Lake City, UT 84112, USA
anil.virkar@m.cc.utah.edu

Raymond A. Cutler
Ceramatec, Inc.
2425 S. 900 W.
Salt Lake City, UT 84119, USA
cutler@ceramatec.com

Chapter 6
High-Thermal-Conductivity SiC
and Applications

J.S. Goela
Rohm and Haas Advanced Materials
185 New Boston Street
Woburn, MA 01801, USA
jgoela@rohmhaas.com

N.E. Brese
Rohm and Haas Electronic Materials
272 Buffalo Ave.
Freeport, NY 11520, USA
NBrese@rohmhaas.com

L.E. Burns
Formerly at Rohm and Haas
Advanced Materials
185 New Boston Street
Woburn, MA 01801, USA
leeburns@shore.net

M.A. Pickering
Rohm and Haas Advanced Materials
185 New Boston Street
Woburn, MA 01801, USA
mpickering@rohmhaas.com

Chapter 7
Chemical-Vapor-Deposited Diamond
for High-Heat-Transfer Applications

J.S. Goela
Rohm and Haas Advanced Materials
185 New Boston Street
Woburn, MA 01801, USA
jgoela@rohmhaas.com

J.E. Graebner
Formerly at
TriQuint Optoelectronics
9999 Hamilton Blvd.
Breinigsville, PA 18031, USA
jegraebner@mailaps.org

Chapter 8
Unusually High Thermal Conductivity
in Carbon Nanotubes

Young-Kyun Kwon
Department of Physics
University of California
Berkeley, CA 94720, USA
ykkwon@civet.berkeley.edu

Philip Kim
Department of Physics,
Columbia University
New York, NY 10027, USA
pkim@phys.columbia.edu

1

Lattice Thermal Conduction
Mechanism in Solids

G.P. Srivastava

The theory of lattice thermal conductivity of nonmetallic solids is presented. After discussing the fundamental issues, the kinetic-theory expression for the conductivity, based on the concept of single-mode phonon relaxation time, is developed in some detail, emphasizing the role of phonon dispersion relations and phonon scattering rates. The theory presented contains only one possible adjustable parameter, *viz.* Grüneisen's anharmonic coefficient γ. The simplified intrinsic conductivity expression, within the high-temperature approximation, is used to derive a set of rules for choosing high-thermal-conductivity materials. The theory is extended to provide a discussion on the conductivity of solids in polycrystalline and low-dimensional forms. Thermal conductivity results of quantum wells, quantum wires, and different solid forms of carbon, *viz.* diamond, graphite, graphene, nanotubes, and fullerenes, are presented and discussed.

1.1 Introduction

One of the fundamental properties of solids is their ability to conduct heat. This property is usually quantified in terms of the thermal conductivity coefficient $\mathcal{K}$, which is defined through the macroscopic expression for the rate of heat energy flow per unit area $\mathbf{Q}$ normal to the gradient ∇T

$$\mathbf{Q} = -\mathcal{K} \, \nabla T. \tag{1.1}$$

Understanding and controlling the thermal conductivity $\mathcal{K}$ of semiconductors plays an important part in the design of power-dissipating devices. For example, power transistors, solar cells under strong sunlight, diodes, transistors, and semiconductor lasers sustain large internal power dissipation, and a high thermal conductivity of the device material can help transfer this energy to a heat sink. On the other hand, a low thermal conductivity of semiconductor alloys helps increase the figure of merit of thermoelectric devices.

In nonmetals heat is conducted by phonons, quanta of atomic vibrational modes. The thermal conductivity of a hypothetical crystal is infinite at all temperatures if it is considered to be infinitely large, is isotopically pure, has no imperfections, and is characterized by purely harmonic atomic vibrations. Within the pure harmonic limit a phonon is infinitely long-lived, characterized with its frequency $\omega(\mathbf{q}s)$ for wave vector $\mathbf{q}$ and polarization index s (longitudinal L or transverse T). Thus, on the application of a temperature gradient, phonons of a purely harmonic crystal would transport all heat from the hot end to the cold end. In other words, the thermal conductivity of a purely harmonic crystal would be infinite at all temperatures. However, real solids are of finite size, contain defects, and exhibit anharmonicity in atomic vibrations. These realities limit the lifetime of phonons, rendering finite values of thermal conductivity. Experimental measurements indicate strong temperature dependence of the conductivity.

Intrinsic phonon-phonon interactions, caused by anharmonicity at finite temperatures, are inelastic in nature. This makes the concept of phonon lifetime an intrinsically difficult, if not impossible, concept to comprehend and thus evaluate theoretically. Thus, it is not usually possible to obtain an exact expression for $\mathcal{K}$. In this chapter we will discuss the progress made toward developing plausible theoretical models for lattice thermal conduction mechanisms in nonmetallic solids. It will be pointed out that whatever theory is adopted for deriving an expression for the thermal conductivity of a nonmetallic solid, its numerical evaluation requires an accurate knowledge of two essential inputs: (1) phonon-dispersion relation, and (2) relevant phonon-scattering mechanisms (to construct the phonon-scattering operator or to derive the phonon relaxation time). After a brief discussion of these aspects, we will follow a simple relaxation-time scheme, based on an isotropic continuum model, to discuss the theory of thermal conduction in crystalline, polycrystalline, and low-dimensional forms of nonmetals. The high-temperature expression for the conductivity will be used to derive a set of rules for choosing high-thermal-conductivity materials. Thermal conductivity results for the various solid forms of carbon, *viz.* diamond, graphite, graphene, nanotubes, and fullerene, will be presented and discussed.

1.2 Theory of Thermal Conductivity

Let us consider a crystal with N_0 unit cells, each of volume Ω. Let us also identify a phonon with its wave vector $\mathbf{q}$, polarization index s, frequency $\omega(\mathbf{q}s)$, and group velocity $c_s(\mathbf{q})$. The heat current $\mathbf{Q}$ can be expressed by including contributions from phonons in all possible modes

$$\mathbf{Q} = \frac{1}{N_0\Omega} \sum_{\mathbf{q}s} \hbar\omega(\mathbf{q}s)n_{\mathbf{q}s}c_s(\mathbf{q}). \qquad (1.2)$$

The quantity $n_{\mathbf{q}s}$, which is explained later, assumes its equilibrium value $\bar{n}_{\mathbf{q}s}$ characterized by the crystal temperature T. In the presence of a temperature gradient across the crystal we can express

$$n_{\mathbf{q}s} = \bar{n}_{\mathbf{q}s} + \delta n_{\mathbf{q}s}, \tag{1.3}$$

where $\delta n_{\mathbf{q}s}$ indicates deviation from the equilibrium value. Clearly, then, the heat current is governed by $\delta n_{\mathbf{q}s}$, so that Eq. (1.2) can be reexpressed as

$$\mathbf{Q} = \frac{1}{N_0\Omega} \sum_{\mathbf{q}s} \hbar\omega(\mathbf{q}s)\delta n_{\mathbf{q}s} c_s(\mathbf{q}). \tag{1.4}$$

The deviation quantity $\delta n_{\mathbf{q}s}$, which is significantly controlled by crystal anharmonicity, particularly at high temperatures, is in general unknown.

Microscopic theories of lattice thermal conductivity attempt to address the quantity $\delta n_{\mathbf{q}s}$. Two fundamentally different approaches have been developed: (1) linear-response methods based on the Green-Kubo formalism and (2) methods based on solving the phonon Bolzmann equation. A detailed discussion of these theoretical methods can be found in the book by Srivastava [1]. Here we will briefly outline the fundamental concepts underlying these approaches.

1.2.1 Green-Kubo Linear-Response Theory

The Green-Kubo formalism is rooted in quantum statistics. It begins by expressing Eq. (1.4) as

$$\mathcal{K} = \frac{k_B T^2 N_0\Omega}{3} \Re \int_0^\infty \langle \mathbf{Q}(0) \cdot \mathbf{Q}(t) \rangle \mathrm{d}t \tag{1.5}$$

$$= \frac{\hbar^2}{3N_0\Omega k_B T^2} \Re \int_0^\infty \mathrm{d}t \sum_{\mathbf{q}s\mathbf{q}'s'} \omega(\mathbf{q}s)\omega(\mathbf{q}'s')\mathbf{c}_s(\mathbf{q}) \cdot \mathbf{c}'_s(\mathbf{q}')\mathcal{C}_{\mathbf{q}s\mathbf{q}'s'}(t), \tag{1.6}$$

where

$$\mathcal{G}(t) \equiv \langle \delta n_{\mathbf{q}s}(0)\delta n_{\mathbf{q}'s'}(t) \rangle \equiv \mathcal{C}_{\mathbf{q}s\mathbf{q}'s'}(t) \tag{1.7}$$

is a *correlation function*. The quantity $n_{\mathbf{q}s}$ is regarded as the number-density operator for phonons in mode $\mathbf{q}s$ in the Heisenberg representation:

$$n_{\mathbf{q}s}(t) = a^\dagger_{\mathbf{q}s}(t)a_{\mathbf{q}s}(t), \tag{1.8}$$

where $a^\dagger_{\mathbf{q}s}$ and $a_{\mathbf{q}s}$ are the creation and annihilation operators, respectively.

As $\delta n_{\mathbf{q}s}$ is generally unknown, an exact evaluation of $\mathcal{G}$ is not possible. Approximate expressions for $\mathcal{G}$ can be derived by employing several theoretical techniques, such as the Zwangiz–Mori projection operator method, double-time Green function method, and imaginary-time Green function method. The first two of these methods have been described in the book by Srivastava [1], to which the interested reader is referred for details.

1.2.2 Variational Principles

In applying variational principles for deriving approximate expressions for $\mathcal{K}$, the quantity $n_{\mathbf{q}s}$ is considered as a distribution function $n_{\mathbf{q}s}(\mathbf{r}, t)$ that measures the occupation number of phonons ($\mathbf{q}s$) in the neighborhood of a point $\mathbf{r}$ in space at time t. In the absence of an external temperature gradient, the thermal average of the distribution function is given by the Bose-Einstein expression

$$\bar{n}_{\mathbf{q}s} = \frac{1}{\exp[\hbar\omega(\mathbf{q}s)/k_B T] - 1}. \tag{1.9}$$

In the steady state of heat flow through a crystal, the total time rate of change of the distribution function $n_{\mathbf{q}s}(\mathbf{r}, t)$ satisfies the Boltzmann equation

$$\left.\frac{\partial n_{\mathbf{q}s}}{\partial t}\right|_{\text{diff}} + \left.\frac{\partial n_{\mathbf{q}s}}{\partial t}\right|_{\text{scatt}} = 0, \tag{1.10}$$

where the first term represents *diffusion* (i.e., variation from point to point) of $n_{\mathbf{q}s}(\mathbf{r}, t)$ through the solid, and the second term represents the rate of change of $n_{\mathbf{q}s}(\mathbf{r}, t)$ due to possible phonon-scattering processes. Noting that in equilibrium $\partial\bar{n}_{\mathbf{q}s}/\partial t = 0$, a canonical form of the linearized phonon Boltzmann equation reads

$$-\mathbf{c}_s(\mathbf{q}) \cdot \nabla T \frac{\partial \bar{n}_{\mathbf{q}s}}{\partial T} = -\left.\frac{\partial n_{\mathbf{q}s}}{\partial t}\right|_{\text{scatt}}. \tag{1.11}$$

This form of the phonon Bolzmann equation can be written in a standard form

$$X_{\mathbf{q}}^s = \sum_{\mathbf{q}'s'} P_{\mathbf{q}\mathbf{q}'}^{ss'} \psi_{\mathbf{q}'}^{s'}, \tag{1.12}$$

where $X_{\mathbf{q}}^s$ is a measure of inhomogeneity caused by the application the of the temperature gradient, $P_{\mathbf{q}\mathbf{q}'}^{ss'}$ is an element of the phonon-scattering operator, and $\psi_{\mathbf{q}}^s \equiv \psi_{\mathbf{q}s}$ is a function that measures the deviation quantity $\delta n_{\mathbf{q}s}$ defined as follows

$$n_{\mathbf{q}s} = \frac{1}{\exp[\hbar\omega(\mathbf{q}s)/k_B T - \psi_{\mathbf{q}s}] - 1} \tag{1.13}$$

$$\simeq \bar{n}_{\mathbf{q}s} + \psi_{\mathbf{q}s}\bar{n}_{\mathbf{q}s}(\bar{n}_{\mathbf{q}s} + 1). \tag{1.14}$$

Using Eqs. (1.4) and (1.12), the following expression for the thermal conductivity can be obtained

$$\mathcal{K} = \frac{k_B T^2}{N_0 \Omega \mid \nabla T \mid^2} \sum_{\mathbf{q}s} \psi_{\mathbf{q}}^s X_{\mathbf{q}}^s. \tag{1.15}$$

This expression cannot be evaluated exactly, as the anharmonic contribution to the deviation function $\psi_{\mathbf{q}s}$ is generally unknown. An effort to express ψ in terms of the inverse scattering operator P^{-1} would remain unsuccessful, as the full set of eigenvalues and eigenvectors of the operator P is not known [2]. Obtaining an approximation for Eq. (1.15) thus becomes a cherished topic of the variational method.

In a simple variational approach the deviation function is expressed as

$$\psi_{\mathbf{q}s} = \phi_{\mathbf{q}s} + \delta\phi_{\mathbf{q}s} \tag{1.16}$$

and the semidefinite property of the operator P is expressed as

$$(\delta\phi, P\delta\phi) \geq 0. \tag{1.17}$$

A simple form of variational trial function $\phi_{\mathbf{q}s}$ can be chosen as

$$\phi_{\mathbf{q}s} = \mathbf{q} \cdot \mathbf{u}, \tag{1.18}$$

where $\mathbf{u}$ is some constant vector parallel to the applied temperature gradient. This choice for a variational trial function has been made from the momentum-conserving property of anharmonic phonon normal (N) processes [3, 4] (also see Sect. 1.4.2). With this choice of the trial function a lower bound $\mathcal{K}_0^<$ for the exact conductivity coefficient $\mathcal{K}$ can be derived [5, 6]: $\mathcal{K}_0^< \leq \mathcal{K}$,

$$\mathcal{K}_0^< = \frac{(\phi, \mathbf{X})}{(\phi, P\phi)^2}, \tag{1.19}$$

where $(\mathbf{f}, \mathbf{g}) = \sum_{\mathbf{q}s} f_{\mathbf{q}s} g_{\mathbf{q}s}$ is implied. The Ziman bound $\mathcal{K}_0^<$ can be improved, i.e., brought closer to $\mathcal{K}$, by employing a more general trial function [7], such as one made as a linear combination of a few simple trial functions in powers of q: $\phi = \sum_{n=1}^N \alpha_n \phi_n$.

The ubiquitous simple form of the variational principle just described can be extended to take the form of *complementary variational principles* [8]. To develop such principles the phonon scattering operator P is split in the form $P = \mathcal{L} + \mathcal{T}^*\mathcal{T}$, such that $\mathcal{L}^{-1}$ exists and $\mathcal{T}^*$ is the conjugate of $\mathcal{T}$. Using the split form of P the phonon Boltzmann equation in Eq. (1.12) can be expressed in a canonical Euler-Lagrange form, from which an upper and a lower bound on $\mathcal{K}$ can be derived. As an alternative to using the Euler-Lagrange variational principles, the upper and lower bounds on $\mathcal{K}$ can also be derived by applying Schwarz's inequality based on the positive semidefinite nature of P and $(P\mathcal{L}^{-1} - \hat{I})$: $(\mathbf{f}, P\mathbf{f}) \geq 0$ and $(\mathbf{f}, (P\mathcal{L}^{-1} - \hat{I})\mathbf{f}) \geq 0$ for any admissible vector function $\mathbf{f}$. Using these ideas, monotonically convergent sequences of lower bounds $\{\mathcal{K}_m^<\}, m = 0, 1, 2, \ldots$ [9] and upper bounds $\{\mathcal{K}_n^>\}, n = 1, 3, 5, \ldots$ [10] for thermal conductivity can be derived.

While the derivation and application of the method of complementary variational principles are somewhat involved, it is easy to appreciate their achievement. Consider a pair of complementary bounds: an upper bound $\mathcal{K}_n^>$ and a lower bound $\mathcal{K}_m^<$, with suitably chosen large values of m and n. From these two bounds we can define a narrow window $\Delta_{m,n} = \mathcal{K}_n^> - \mathcal{K}_m^<$ within which the theoretical estimate of exact conductivity $\mathcal{K}$ must lie. For details on the applications of the complementary variational principles, interested readers are referred to chapter 5 in the the book by Srivastava [1] and references therein.

1.2.3 Relaxation-Time Approaches

The difficulty in deriving an (approximate) expression for lattice thermal conductivity can be appreciated from the discussion provided in the previous two subsections. Relaxation-time approaches provide simple alternatives to the variational approaches for deriving expressions for the conductivity. In general, due to their inelastic nature, anharmonic phonon interactions are not amenable to a relaxation-time picture [11]. But for simplicity of understanding the problem it is useful to introduce the concept of anharmonic phonon relaxation time. Starting from the phonon Boltzmann equation, the quantity $-\partial n_{\mathbf{q}s}/\partial t|_{\text{scatt}}$ in Eq. (1.11), or $P\psi$ in Eq. (1.12), can be expressed using the concept of a relaxation-time $\tau_{\mathbf{q}s}$ for a phonon in mode $\mathbf{q}s$. It is assumed that on application of a temperature gradient each phonon in mode $\mathbf{q}s$ transports heat during its lifetime (i.e., before it is annihilated due to scattering events). Important contributions to the theory of thermal conductivity based on relaxation-time approaches have been made, among others, by Debye [12], Klemens [13], Callaway [14], Simons [15], and Srivastava [1, 16].

The simplest of relaxation-time approaches is the so-called *single-mode relaxation-time* method. In this picture it is assumed that while phonons in mode $\mathbf{q}s$ have been driven out of their equilibrium distribution on application of a temperature gradient and transport heat for the duration of their lifetime $\tau_{\mathbf{q}s}$, phonons in all other modes remain in their thermal equilibrium. In the language of the previous subsection, this means that while $\psi_{\mathbf{q}s} \neq 0$, $\psi_{\mathbf{q}'s'} = 0$ for $\mathbf{q}'s' \neq \mathbf{q}s$. With this restriction we can represent the scattering operator P by its diagonal part only:

$$\sum_{\mathbf{q}'s'} P_{\mathbf{q}\mathbf{q}'}^{ss'} \psi_{\mathbf{q}'}^{s'} \simeq P_{\mathbf{q}\mathbf{q}}^{ss} \psi_{\mathbf{q}}^{s}. \tag{1.20}$$

This allows the right-hand side of the Boltzmann equation in Eqs. (1.11) and (1.12) to be simplified to

$$-\frac{\partial n_{\mathbf{q}s}}{\partial t}\bigg|_{\text{scatt}} = \frac{n_{\mathbf{q}s} - \bar{n}_{\mathbf{q}s}}{\tau_{\mathbf{q}s}} = \frac{\bar{n}_{\mathbf{q}s}(\bar{n}_{\mathbf{q}s} + 1)\psi_{\mathbf{q}s}}{\tau_{\mathbf{q}s}} = P_{\mathbf{q}\mathbf{q}}^{ss} \psi_{\mathbf{q}}^{s}. \tag{1.21}$$

The single-mode relaxation-time $\tau_{\mathbf{q}s}$ is thus defined from the relation

$$\tau_{\mathbf{q}s}^{-1} = \frac{P_{\mathbf{q}\mathbf{q}}^{ss}}{\bar{n}_{\mathbf{q}s}(\bar{n}_{\mathbf{q}s} + 1)}. \tag{1.22}$$

Using Eqs. (1.12) and (1.15), the single-mode relaxation-time expression for thermal conductivity $\mathcal{K}_{smrt}$ becomes

$$\mathcal{K}_{smrt} = \frac{\hbar^2}{3N_0 \Omega k_B T^2} \sum_{qs} c_s^2(\mathbf{q}) \omega^2(\mathbf{q}s) \tau_{\mathbf{q}s} \bar{n}_{\mathbf{q}s}(\bar{n}_{\mathbf{q}s} + 1). \qquad (1.23)$$

This expression can be viewed as the kinetic theory result

$$\mathcal{K} = \frac{1}{3} C_v^{sp} \bar{c}^2 \bar{\tau}, \qquad (1.24)$$

with C_v^{sp} as the phonon specific heat (heat capacity per unit volume), $\bar{c}$ as average phonon speed, and $\bar{\tau}$ as average phonon relaxation time. The single-mode relaxation-time expression in Eq. (1.23) is sometimes known as the Debye conductivity expression, as it was first used by Debye [12].

A significant improvement over the single-mode relaxation-time approach was made by Callaway [14], who included the special role played by the momentum conservation in anharmonic phonon interaction. In the language of phonon-scattering operator P, the extra contribution to the phonon relaxation time over the single-mode result is that due to the off-diagonal part of the operator corresponding to the momentum-conserving normal (N) processes. The Callaway expression for the conductivity is of the form given in Eq. (1.23), but with the single-mode relaxation-time τ replaced by an *effective relaxation-time* τ_C

$$\tau_C = \tau(1 + \beta_C/\tau_N), \qquad (1.25)$$

where the parameter β_C is a function of the single-mode relaxation-time τ and a contribution τ_N from anharmonic N-processes. The resulting conductivity expression can be expressed as

$$\mathcal{K}_C = \mathcal{K}_{smrt} + \mathcal{K}_{N\text{-drift}}. \qquad (1.26)$$

The contribution from the N-drift term can be significantly important for pure materials.

An attempt to incorporate the role of off-diagonal anharmonic momentum nonconserving (umklapp, or U) processes was made by Srivastava [16]. The resulting model conductivity expression is similar to the Callaway expression, but with τ_C replaced by τ_S, where

$$\tau_S = \tau_m(1 + \beta_S \tau_N) \qquad (1.27)$$

with β_S including the effect of τ_m. Here τ_m includes a modification of the single-mode relaxation-time τ arising from the contribution of U-processes to the off-diagonal part of the operator P. Clearly, $\tau_S = \tau_C$ in the absense of such an attempt (i.e., when $\tau_m = \tau$).

1.3 Phonon-Dispersion Relations

From the discussion in the previous section it is clear that whatever level of the theory of lattice thermal conductivity we decide to use, evaluation of $\mathcal{K}$ requires knowledge of phonon-dispersion relation $\omega = \omega(\mathbf{q}s)$, phonon relaxation-time $\tau(\mathbf{q}s)$, and a scheme for performing summation over phonon wave vectors $\mathbf{q}$ and polarization s of the expression in Eq. (1.23). Calculation of the phonon-dispersion relation is a research topic in its own right, known as lattice dynamics, and is beyond the scope of this chapter. Interested readers are referred to the books in Refs. [1, 17]. From a knowledge of $\omega(\mathbf{q}s)$ for a given polarization index s the sum over $\mathbf{q}$ can be replaced by an integral

$$\sum_{\mathbf{q}} \rightarrow \int g(\omega)\,d\omega, \qquad (1.28)$$

where $g(\omega)$ is the density of phonon states at the frequency ω. In general, the density of states is inversely proportional to the magnitude of the phonon group velocity

$$g(\omega) \propto \frac{1}{|\nabla_{\mathbf{q}}\omega|}. \qquad (1.29)$$

For a proper calculation of the density of states $g(\omega)$ it is essential to obtain numerical values of the phonon-dispersion relation $\omega(\mathbf{q}s)$ for a large number of phonon wave vectors $\mathbf{q}$ within the irreducible part of the Brillouin zone of the solid under consideration. We will not, however, discuss this further in the present context.

With a view to restricting our discussion to high-thermal-conductivity materials, in this section we will examine the main features of the phonon-dispersion relation in diamond and aluminium nitride as examples of three-dimensional systems, graphite as an example of *quasi*–two-dimensional system, and carbon nanotubes as examples of *quasi*–one-dimensional systems. We will also present a simplified version of the dispersion relation and density of states of these systems within the continuum limit for crystal structure.

1.3.1 Three-Dimensional Materials

(i) *Diamond.* The diamond crystal structure can be constructed from a consideration of the face-centered cubic lattice and by assigning each lattice point two carbon atoms at a relative separation of $a(1/4, 1/4, 1/4)$ from each other, where a is the cubic lattice constant. The atomic positions of a solid with a closely related structure, the zincblende structure, within the conventional cubic unit cell of length a, is shown in Fig. 1.1. The diamond structure can be identified with the zincblende structure when the two basis atoms are of the same species.

The phonon-dispersion relation and the density of states for diamond are shown in Fig. 1.2. The dispersion results are plotted along the three principal

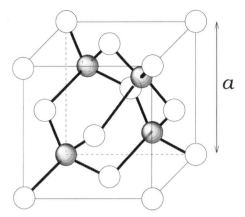

Fig. 1.1. Atomic positions in a solid with the zincblende structure, shown in the conventional cubic unit cell of size a.

symmetry directions: from Γ to X along [100], from Γ to K and extended up to X along [110], and from Γ to L along [111]. Due to two carbon atoms per primitive unit cell, there are six phonon branches. With increasing energy near the Brillouin zone center (Γ) these are the T_1A (slow transverse acoustic), T_2A (fast transverse acoustic), LA (longitudinal acoustic), T_2O (fast transverse optical), T_1O (slow transverse optical) and LO (longitudinal optical) branches. The density of states shows a few characteristic peaks, corresponding to flatness of the dispersion curves for the various polarization branches.

(ii) *β-AlN.* Aluminium nitride can assume two crystal phases: zincblende and wurtzite, known as *β*-AlN and *α*-AlN, respectively. In the zincblende

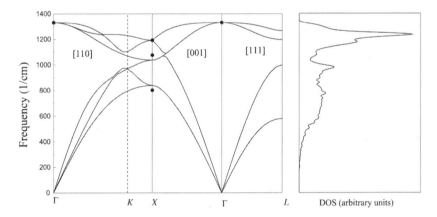

Fig. 1.2. Phonon-dispersion curves and density of states for diamond. The results are obtained from the application of a bond charge model [18]. Experimental measurements are indicated by filled circles.

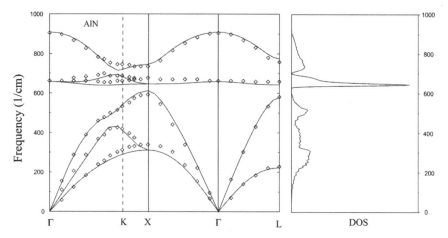

Fig. 1.3. Phonon-dispersion curves and density of states for AlN in the zincblende phase (β-AlN). Solid curves are obtained from calculations employing a bond charge model, and empty diamonds indicate experimental results. Taken from [19].

phase the cubic lattice constant is 4.38 Å. The phonon-dispersion curves for zincblende AlN, together with the density-of-states curve, are shown in Fig. 1.3. There are two significant differences between the dispersion curves for diamond and AlN. First, the TO branch in AlN is very flat, leading to a sharp peak in the density of states. Second, there is a large LO − TO splitting at the zone center (the Γ point) for AlN. This arises due to the ionic nature of AlN.

(iii) α-*AlN*. The crystal structure of the wurtzite phase of AlN can be constructed from the hexagonal lattice with a basis of four atoms such as: Al at $(0,0,0)$, $(2a/3, a/3, c/2)$ and N at $(0,0,u)$, $(2a/3, a/3, c/2 + u)$. For α-AlN the hexagonal lattice constants are $a = 3.11$ Å and $c = \sqrt{(8/3)}a$, and the internal parameter is $u = 0.382$. Although each atom is tetrahedrally bonded to four neighbors of another species in both the zincblende and wurtzite structures, the connectivity of covalent bonds is different in the two structures. The crystal structure and the phonon-dispersion curves and the density of states for the α phase are shown in Figs. 1.4 and 1.5, respectively.

As a result of the geometrical differences between the β and α phases, there are a few differences in the phonon spectra and the density of states for the two phases. The four atoms within the wurtzite primitive unit cell result in three acoustic and nine optical branches. The lowest three optical branches are found to lie in the acoustic range obtained for the zincblende phase. The other six optical branches are well separated from the acoustic and the lower-lying optical branches. Thus there is an optical-optical gap in the phonon spectrum for α-AlN, as opposed to the optical-acoustic gap for β-AlN. This occurs because the extent of the Brillouin zone in the wurtzite phase along the [111] direction is only half of that in the zincblende phase. The density of

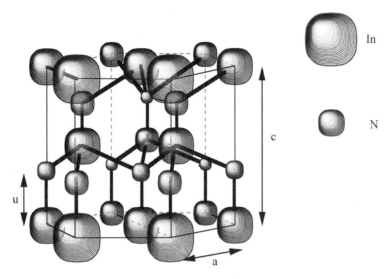

Fig. 1.4. Wurtzite crystal structure.

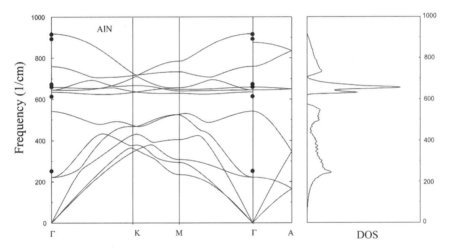

Fig. 1.5. Phonon-dispersion curves and density of states for AlN in the wurtzite phase (α-AlN). Solid curves are obtained from calculations employing a bond charge model, and filled circles indicate experimental results. Taken from [19].

states for α-AlN shows the development of a small but sharp peak just below the large peak in the lower part of the optical range for the zincblende phase.

1.3.2 Graphite, Graphene, and Nanotubes

(i) *Graphite.* The graphite structure is characterized by a basis of four carbon atoms assigned to each point of a simple hexagonal lattice. The distribution of carbon atoms can be visualized in the form of atomic planes (the so-called basal planes). The basal atomic planes perpendicular to the c-axis have a honeycomb arrangement. The interplanar separation is 2.36 times the nearest-neighbor interatomic distance (1.42 Å) in a basal plane, indicating much weaker interlayer bonding.

The primitive translation vectors in a basal plane of graphite can be chosen as shown in Fig. 1.6. Corresponding to four atoms per primitive cell, there are 12 phonon branches for graphite, shown in Fig. 1.7. However, these branches are rather strangely ordered and show anomalous dispersion [20]. In a basal-plane direction there are three acoustic branches. While the LA and fast TA branches show normal dispersion, the slow TA branch (also called the bending mode branch) shows an anomalous dispersion $\omega \propto q^2$ for low q-values and a linear behavior $\omega \propto q$ for larger values of q. There is also a very low-lying TO branch at the zone center which shows a dispersion behavior similar to that of the slow TA branch. The frequencies of the in-plane TA, LA, and LO branches extend up to about 25 THz, 32 THz, and 47 THz, respectively. Along the interplanar direction the TA and LA branches are very low-lying (below $\nu_c(\mathrm{LA}) = (\omega_c(\mathrm{LA})/2\pi) \simeq 4$ THz) and get folded into dispersionless TO and LO branches, respectively.

(ii) *Graphene.* A single graphite plane is a graphene sheet. This hypothetical form of carbon, therefore, contains two carbon atoms per unit cell, leading to 6 phonon branches, as shown in Fig. 1.8(a) [21]. The three optical branches correspond to one out-of-plane mode and two in-plane modes. Near the zone center, with increasing energy the three acoustic branches correspond to an out-of-plane mode, an in-plane tangential (bond-bending) mode, and an in-plane radial (bond-stretching) mode, respectively. The out-of-plane (transverse) mode shows a q^2 dispersion, similar to that of the slow TA mode in graphite. The density of acoustic modes in the graphene sheet is a constant, as seen in Fig 1.8(b).

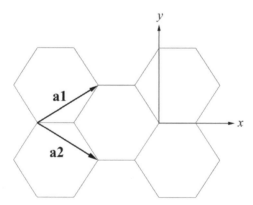

Fig. 1.6. The basal plane of graphite (i.e., a graphene sheet). The primitive translation vectors are indicated.

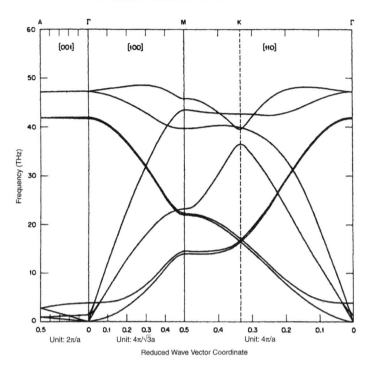

Fig. 1.7. Phonon-dispersion curves for the three-dimensional form of graphite. Taken from [20] with permission.

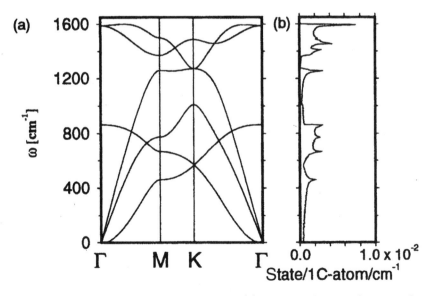

Fig. 1.8. (a) Phonon-dispersion curves and (b) density of states for a graphene sheet. Taken from [21] with permission.

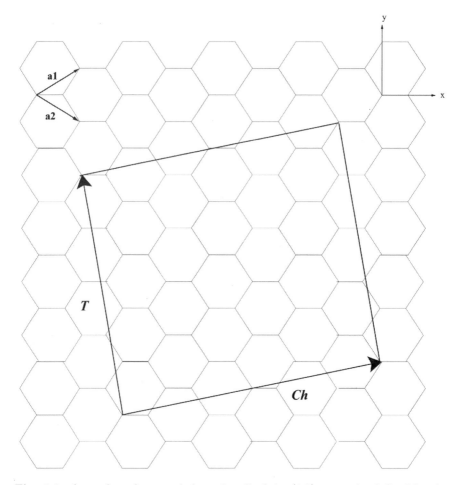

Fig. 1.9. A graphene layer and the unit cell of the (4,2) nanotube defined by the chiral vector $\mathbf{C}_h$ and the translation vector $\mathbf{T}$. There are 56 atoms in the nano-unit cell.

(iii) *Nanotubes.* A carbon nanotube is made by rolling up a graphene sheet. The structure of a single-wall nanotube can be specified by its chiral vector $\mathbf{C}_h$ and its translation vector $\mathbf{T}$. These vectors can be expressed as suitable linear combinations of the primitive translation vectors $\mathbf{a}_1$ and $\mathbf{a}_2$ of the graphene sheet. In particular, $\mathbf{C}_h = n\mathbf{a}_1 + m\mathbf{a}_2 \equiv (n, m)$. The diameter D of the tube is $D = |\mathbf{C}_h|/\pi$. An armchair nanotube corresponds to $\mathbf{C}_h = (n, n)$. The lattice translation vector parallel to the tube axis $\mathbf{T}$ is normal to the chiral vector $\mathbf{C}_h$: $\mathbf{C}_h \cdot \mathbf{T} = 0$. Fig. 1.9 shows the $\mathbf{C}_h$ and $\mathbf{T}$ vectors for the (4,2) nanotube.

Figure 1.10 shows the phonon-dispersion curves and the density of states for the (10,10) carbon nanotube [21]. There are 40 carbon atoms in the unit cell for the (10,10) nanotube, giving rise to 120 phonon branches. In accordance

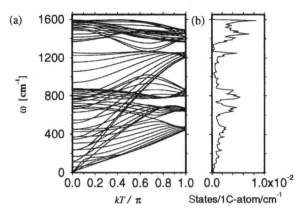

Fig. 1.10. (a) Phonon-dispersion curves and (b) density of states (b) for the (10,10) carbon nanotube. Taken from [21] with permission.

with the D_{10h} point group symmetry, 12 branches are non-degenerate and 54 branches are doubly degenerate, thus there are only 66 distinct branches. A nanotube is characterized by four acoustic branches: with increasing energy these are two (degenerate) TA modes, a twisting mode (related to a rotation around the tube axis), and the longitudinal mode (in the direction of the tube axis), respectively [21, 22]. In a radial direction, the phonon wave vectors are quantized due to the periodic boundary conditions imposed by the cylindrical symmetry. The density of acoustic modes in the nanotube is close to that of a graphene layer, except for two noticeable differences. First, compared to the graphene layer, there are a few extra small peaks in the density of states for the nanotube, indicative of its quasi–one-dimensional nature. Second, whereas the density of states is finite for the graphene layer, it goes to zero for the nanotube, as $\omega \to 0$. This difference is due to the fact that $\omega \propto q^2$ for the graphene layer and $\omega \propto q$ for the nanotube in the limit $\omega \to 0$.

1.3.3 Debye's Isotropic Continuum Model

Implementation of a realistic phonon-dispersion relation and density of states in a calculation of thermal conductivity would obviously require much effort and time. Most calculations of the conductivity have, therefore, been made by considering simplified forms of the dispersion relation and density of states. One particularly simple scheme is provided by the consideration of the *isotropic continuum model* and a Debye cutoff scheme. In this scheme the realistic Brillouin zone for a three-dimensional cubic system is replaced by a Debye sphere of radius q_D and the phonon-dispersion relation is simplified to $\omega(\mathbf{q}s) = qc_s$ for all directions of $\mathbf{q}$. This relation is certainly true in the long wavelength (small q) limit. An inspection of the dispersion curves in Figs. 1.2 and 1.3 suggests that up to approximately 60 percent of both the LA and TA

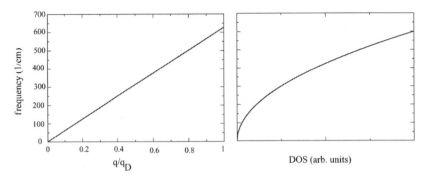

Fig. 1.11. Phonon-dispersion curve and density of states for a single polarization branch within the Debye model of a three-dimensional isotropic continuum.

branches can be reasonably well described by the linear dispersion relation. The size of the Debye sphere, determined by ensuring that it contains the correct number of acoustic phonon modes ($3N$ modes for a crystal with N atoms), is given by

$$q_{\mathrm{D}} = \left(\frac{6\pi^2 N}{N_0 \Omega} \right)^{1/3}. \tag{1.30}$$

The density of states within the isotropic continuum model is given by the following simple expression

$$g(\omega_s)\big|_{\text{isotropic continuum}} = \frac{N_0 \Omega}{2\pi^2} \frac{\omega_s^2}{c_s^3}. \tag{1.31}$$

Figure 1.11 shows the phonon-dispersion curve and the density of states within the Debye model for a single polarization branch. The density of states increases quadratically with increasing phonon frequency but in contrast to the real situation (*cf.* Fig. 1.2) there are no characteristic peaks.

These considerations can also be applied to deriving expressions for the phonon density of states for two- and one-dimensional systems. It can be shown that the ω^2 variation of the density of states for the three-dimensional isotropic continuum changes to ω and ω^0 (i.e., a constant) for two- and one-dimensional systems, respectively.

1.4 Phonon Relaxation Times

Finite sample size, static imperfections, alloying and inhomogeneity, and anharmonicity in crystal potential provide the main phonon-scattering sources in nonmetallic solids. Each mechanism acts in limiting the lifetime of phonons. Although anharmonicity gives rise to *intrinsic relaxation time*, other mechanisms produce *extrinsic relaxation times*. We will reproduce expressions for phonon relaxation times due to some of these mechanisms in bulk (three-dimensional) crystals.

1.4.1 Extrinsic Relaxation Times

(i) *Boundary Scattering.* At low temperatures phonons acquire long wavelengths and the main source of their scattering is the sample size. The phonon relaxation rate due to boundary scattering can be expressed as

$$\tau_{\mathbf{q}s}^{-1}(\mathrm{bs}) = c_s/L. \tag{1.32}$$

Here L represents an effective boundary mean free path and can be expressed as [6] $L = L_0(1+p)/(1-p)$, with L_0 representing the geometrical mean free path and $p \neq 0$ representing surface nonspecularity. For a cylindrical shape $L_0 = D$, the circular cross section. For a square cross section $L_0 \simeq 1.12d$, for side length d.

(ii) *Scattering from Static Point Imperfections.* Static point imperfections in solids, such as isotopes, substitutional impurities with different masses, impurities causing changes in atomic force constants, and single and aggregate vacancy defects, can strongly scatter phonons. For phonons with longer wavelengths compared with the imperfection size, the scattering is essentially of the Rayleigh type, that is, increases as the fourth power of frequency. A general expression for the relaxation rate of phonons from point imperfections is [1, 23]

$$\tau_{\mathbf{q}s}^{-1}(\mathrm{md}) = \frac{\Gamma_{\mathrm{md}}\Omega}{4\pi\bar{c}^3}\omega^4(\mathbf{q}s) \tag{1.33}$$

$$= \frac{3\pi}{2}\Gamma_{\mathrm{md}}\frac{\omega^4}{\omega_{\mathrm{D}}^3}, \tag{1.34}$$

with Γ_{md} determined from the nature of imperfection and $3\bar{c}^{-3} = \sum_s c_s^{-3}$. For a general case [23]

$$\Gamma_{\mathrm{md}} = \sum_i f_i \left[\left(1 - \frac{M_i}{\bar{M}}\right)^2 + 2\left(\frac{\Delta g_i}{g} - 6.4\gamma\frac{\Delta\delta_i}{\delta}\right)^2 \right], \tag{1.35}$$

where $\bar{M}$ is the average atomic mass, f_i is the fraction of the unit cells having atomic mass M_i, the fractional size of the imperfection is expressed as $\Delta\delta_i/\delta$, $\Delta g_i/g$ represents the fractional stiffness constant of the nearest-neighbor bonds from the imperfection to the host crystal, and γ is an average anharmonicity of bonds linking the imperfection.

(iii) *Scattering from Imperfection Aggregates, Dislocations, Stacking Faults, and Grain Boundaries.* The frequency dependence of scattering shows departure from the Rayleigh type ω^4 behavior to a ω^2 power law when the phonon wavelength becomes comparable to or smaller than a defect aggregate (see, e.g., [24]). Also, the ω^4 dependence of phonon scattering by point imperfections changes to ω^2 for scattering by dislocations (core region), stacking faults, and grain boundaries, and to ω for scattering by the elastic region of dislocations [6, 23].

1.4.2 Intrinsic Relaxation Times

Atomic vibrations become increasingly anharmonic as the temperature increases, and phonon-phonon interactions become dominantly important in controlling thermal properties of a crystal. As mentioned earlier, describing and quantifying such interactions is an inherently difficult problem. Limiting the crystal anharmonicity to the cubic terms in atomic displacement, we can treat three-phonon interactions using first-order perturbation theory and four-phonon interactions using second-order perturbation theory. In general, the strength of four-phonon interactions is two to three orders of magnitude weaker than the strength of three-phonon interactions [25]. There is a great deal of discussion in the literature regarding frequency and temperature dependence of three-phonon relaxation rates (see, e.g., Herring [26], Klemens [13], Ziman [6], Guthrie [27]). Unfortunately, there is no unanimous agreement in both low- and high-temperature regions. In many works the anharmonic relaxation rate has been expressed as a product of frequency- and temperature-dependent terms: $\tau_{3\mathrm{ph}} \propto f_1(\omega) f_2(T)$, where $f_1(\omega) = A\omega^n$ and $f_2(T) = BT^m$, with exponents n and m chosen differently in different temperature regions. However, a systematic theoretical approach [1, 28, 29] suggests $\tau_{3\mathrm{ph}} \propto f(\omega, T)$, where $f(\omega, T)$ is a rather complicated continuous function over the entire range of frequencies and temperatures.

A three-phonon process can be classified as either class 1 or class 2, as shown in Fig. 1.12. In a class 1 process a phonon $(\mathbf{q}, \omega)$ interacts with another phonon $(\mathbf{q}', \omega')$, and their mutual annihilation creates a third phonon $(\mathbf{q}'', \omega'')$. In a class 2 process an energetic phonon $(\mathbf{q}, \omega)$ decays into two less-energetic phonons $(\mathbf{q}', \omega')$ and $(\mathbf{q}'', \omega'')$. Both processes can take place either in a normal manner (N process, with momentum sum of the two annihilated/created phonons being confined to the first Brillouin zone), or in an umklapp manner (U process, with momentum sum of the two annihilated/created phonons lying beyond the first Brillouin zone and requiring to be *flipped back* to the zone with the help of a reciprocal lattice vector). Obtaining an expression for the cubic anharmonic term in crystal potential is an extremely difficult task. Good progress, however, can be made by treating cubic anharmonicity within an isotropic continuum model. Although there is no meaning of an umklapp process within the continuum model of a solid, Parrott [30] devised a "grafting" scheme for a pseudo-reciprocal lattice vector $\mathbf{G} = 2\mathbf{q}_{\mathrm{D}}\,(\mathbf{q} \pm \mathbf{q}')/(|\mathbf{q} \pm \mathbf{q}'|)$, where the $+$ and $-$ signs refer to class 1 and 2 processes, respectively. The

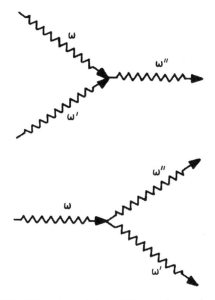

Fig. 1.12. Class 1 and class 2 three-phonon processes.

energy and momentum conservations conditions are:

Class 1 processes:

$$\mathbf{q} + \mathbf{q}' = \mathbf{q}'' + \mathbf{G}$$
$$\omega + \omega' = \omega'', \tag{1.36}$$

Class 2 processes:

$$\mathbf{q} + \mathbf{G} = \mathbf{q}' + \mathbf{q}''$$
$$\omega = \omega' + \omega''. \tag{1.37}$$

(i) *Interactions Involving Acoustic Phonons.* Phonon-phonon interactions can be studied by applying Fermi's golden rule formula, based on first-order time-dependent perturbation theory. Application of this formula within an isotropic continuum anharmonic model for crystal potential leads to the following expression for the single-mode relaxation rate due to three-phonon interactions involving acoustic phonons [1, 28, 29]:

$$\tau_{\mathbf{q}s}^{-1}(3\text{ ph}) = \frac{\pi\hbar}{4\rho^3 N_0 \Omega} \sum_{\mathbf{q}'s'\mathbf{q}''s''} |A_{\mathbf{q}\mathbf{q}'\mathbf{q}''}^{ss's''}|^2 \frac{qq'q''}{c_s c_{s'} c_{s''}^2} \delta_{\mathbf{q}+\mathbf{q}'+\mathbf{q}'',\mathbf{G}}$$

$$\times \left\{ \frac{\bar{n}_{\mathbf{q}'s'}(\bar{n}_{\mathbf{q}''s''}+1)}{(\bar{n}_{\mathbf{q}s}+1)} \delta(\omega(\mathbf{q}s) + \omega(\mathbf{q}'s') - \omega(\mathbf{q}''s'')) \right.$$

$$\left. + \frac{1}{2} \frac{\bar{n}_{\mathbf{q}'s'}\bar{n}_{\mathbf{q}''s''}}{\bar{n}_{\mathbf{q}s}} \delta(\omega(\mathbf{q}s) - \omega(\mathbf{q}'s') - \omega(\mathbf{q}''s'')) \right\}. \tag{1.38}$$

The quantity $|A_{\mathbf{q}\mathbf{q}'\mathbf{q}''}^{ss's''}|^2$, representing the three-phonon interaction strength, can be expressed in terms of second- and third-order elastic constants. A simplified form of this quantity is [1]

$$|A_{\mathbf{q}\mathbf{q}'\mathbf{q}''}^{ss's''}|^2 = \frac{4\rho^2}{\bar{c}^2}\gamma^2 c_s^2 c_{s'}^2 c_{s''}^2, \tag{1.39}$$

with the mode-averaged Grüneisen constant γ providing a measure of anharmonicity. Within Debye's isotropic continuum model, and using Eq. (1.39), Eq. (1.38) can be written as

$$\tau_{\mathbf{q}s}^{-1}(3\text{ ph}) = \frac{\hbar q_{\mathrm{D}}^5 \gamma^2}{4\pi\rho\bar{c}^2}\sum_{s's''\epsilon} c_s c_{s'}$$

$$\times \left(\int dx' x'^2 x_+'' \{1 - \epsilon + \epsilon(Cx + Dx')\}\frac{\bar{n}_{\mathbf{q}'s'}(\bar{n}_+'' + 1)}{(\bar{n}_{\mathbf{q}s} + 1)}\right.$$

$$\left.+\frac{1}{2}\int dx' x'^2 x_-'' \{1 - \epsilon + \epsilon(Cx - Dx')\}\frac{\bar{n}_{\mathbf{q}'s'}\bar{n}_-''}{\bar{n}_{\mathbf{q}s}}\right), \tag{1.40}$$

where q_{D} is the Debye radius, $x = q/q_{\mathrm{D}}, x' = q'/q_{\mathrm{D}}, x_\pm'' = Cx \pm Dx', C = c_s/c_{s''}, D = c_{s'}/c_{s''}, \bar{n}_\pm'' = \bar{n}(x_\pm''),$ and $\epsilon = 1$ (-1) for N (U) processes. The integration limits over the variable x' are determined from considerations of energy and momentum conservation conditions (cf. Eqs. (1.36) and (1.37)).

From Eq. (1.40) it is clear that the three-phonon relaxation rate is a rather complicated function of temperature and phonon frequency and has the general form $\tau_{\mathbf{q}s}^{-1}(3\text{ ph}) = f(\omega(\mathbf{q}s), T)$. However, it is easy to show the following low-temperature (LT) and high-temperature (HT) behaviors

$$\tau^{-1}(HT) = (\mathcal{A}\omega + \mathcal{B}\omega^2)T, \tag{1.41}$$

$$\tau^{-1}(LT)|_{\mathrm{N}} = \mathcal{C}\omega^m T^{5-m}, \tag{1.42}$$

$$\tau^{-1}(LT)|_{\mathrm{U}} = \mathcal{D}f_1(\omega) + \mathcal{E}f_2(\omega, T)\mathrm{e}^{-\alpha/T}, \tag{1.43}$$

where $m = 1, 2, 3, 4,$ or 5 and $\mathcal{A}, \mathcal{B}, \mathcal{C}, \mathcal{D},$ and $\mathcal{E}$ are constants. The low-temperature behavior for N processes in Eq. (1.42) was first derived by Herring [26]. The exponential term in Eq. (1.43) indicates that U processes "freeze out" at low temperatures, with an appropriate constant α. The function f_2 in Eq. (1.43) has the temperature dependence given in Eq. (1.42). It should be clear that a great deal of freedom may be exercised in employing "empirically chosen" functional forms of $\tau^{-1}(3\text{ ph})$ in both low- and high-temperature regions. This "freedom" has indeed been exercised in most theoretical calculations of thermal conductivity. This practice is particularly exemplified in the work by Guthrie [27].

(ii) *Role of Optical Phonons.* Optical phonons can contribute to the thermal conduction process in two different manners. An optical phonon from a dispersive branch can act as a carrier of heat and contribute to the conductivity. However, as optical phonon branches are in general less dispersive than acoustic branches, their contribution toward conductivity is usually very small. More significantly, optical phonons can interact with acoustic phonons and cause extra contribution to three-phonon relaxation time and hence thermal resistivity.

Let us first consider materials that have simple crystal structures and contain atoms of nearly equal masses, such as Si, Ge, and GaAs, all of which contain two atoms per unit cell. For all these crystals there is little or no energy separation between the acoustic and optical phonon branches. In such cases, the Debye model described in the previous subsection for acoustic phonons can be extended to include optical phonon branches. This can be done by increasing the Debye radius q_D to include both acoustic and optical branches. Rewriting Eq. (1.30) in the form $q_D = (6\pi^2 n/\Omega)^{1/3}$, where Ω is the unit cell volume, for diamond and zincblende structure materials we consider $n = 1$ when considering only acoustic phonon modes (i.e., we consider one atom per unit cell, as required in the original Debye consideration), and we consider $n = 2$ to include optical phonons as well.

For materials with a simple crystal structure but different atomic masses, such as AlSb and GaN, it would be reasonable to assume a flat dispersion relation (i.e., Einstein model) for optical phonons. This model would disallow any heat transport by optical phonons, but interactions of type ac + ac → op would be allowed, provided the mass ratio lies in the range $1 < m_1/m_2 \leq 3$. The approach described previously can be adopted to derive an expression for this interaction (see, e.g., [1]).

For complex crystal structures, that is, with more than 2 atoms per unit cell, there is an increased possibility of acoustic-optical interaction. Group III nitrides in the wurzite structure, with four atoms per unit cell, can be classified as complex-structure materials in the context of the present discussion. Roufosse and Klemens [31] have, however, shown that the strength of three-phonon interaction remains substantially independent of n, the number of atoms in the unit cell.

1.5 Conductivity of Single Crystals

1.5.1 Simplified Conductivity Integral

A simplified expression for lattice thermal conductivity can be obtained by converting the single-mode relaxation-time expression in Eq. (1.23) to an integral form within Debye's isotropic continuum model. The resulting expression is

$$\mathcal{K} = \frac{\hbar^2 q_D^5}{6\pi^2 k_B T^2} \sum_s c_s^4 \int_0^1 dx\, x^4 \tau \bar{n}(\bar{n} + 1), \qquad (1.44)$$

where $x = q/q_D$ and τ and $\bar{n}$ are understood as functions of x, T and polarization s. The total relaxation-time is $\tau_{\mathbf{q}s}^{-1} = \sum_i \tau_{\mathbf{q}s}^{-1}(i)$, with $\tau_{\mathbf{q}s}(i)$ being the relaxation time due to the ith phonon-scattering mechanism. Provided reasonable forms of the phonon-dispersion relation $\omega = \omega(\mathbf{q}s)$ and relaxation time $\tau(\mathbf{q}s)$ are chosen, it should be straightforward to evaluate the thermal conductivity of a crystal.

1.5.2 Temperature Variation of Conductivity

From Eqs. (1.24) and (1.44) it is clear that the temperature variation of the conductivity is governed by the joint temperature variation of the lattice specific heat and phonon relaxation time. It is well known that the lattice specific heat rises as T^3 as very low temperatures and saturates to a constant at high temperatures. The phonon relaxation time is controlled by boundary scattering at very low temperatures with no temperature dependence and by anharmonic interactions at high temperatures. Thus it is obvious that the conductivity of a bulk material will sharply rise as T^3 at very low temperatures and decrease with temperature in the high-temperature limit. The conductivity will thus reach a maximum in the intermediate temperature region. The maximum of the conductivity will largely be goverened by scattering of phonons from defects and impurities. The decrease of conductivity at high temperatures will follow a T^{-1} behavior if only three-phonon processes are operative but will show a decrease slighter stronger than T^{-1} if four-phonon processes start to play a significant role.

1.5.3 High-Thermal-Conductivity Materials

A material with room-temperature value of $\mathcal{K}$ larger than $100\,\mathrm{Wm^{-1}K^{-1}}$ is regarded as a high-thermal-conductivity material. In the high-temperature limit the conductivity expression in Eq. (1.44) can be approximated as

$$\mathcal{K}(\mathrm{HT}) = B\bar{M}\Omega_{\mathrm{at}}^{1/3}\Theta_{\mathrm{D}}^3/(T\gamma^2), \qquad (1.45)$$

where B is a constant and Θ_{D} is Debye's temperature. This result suggests four criteria for choosing high-thermal-conductivity materials: (i) low atomic mass, (ii) strong interatomic bonding, (iii) simple crystal structure, and (iv) low anharmonicity. Conditions (i) and (ii) help increase the quantity $\bar{M}\Theta_{\mathrm{D}}^3$ in Eq. (1.45), condition (iii) means a low number of atoms per unit cell, resulting in fewer optical branches and hence fewer anharmonic interactions, and condition (iv) means reduction in the strength of anharmonic interactions. Consistent with this suggestion, at least 12 semiconductors and insulators can be categorized as high-thermal-conductivity materials. In order of decreasing room-temperature conductivity these are: C (diamond), BN, SiC, BeO, BP, AlN, BeS, BAs, GaN, Si, AlP, and GaP. Table 1.1 lists the room-temperature thermal conductivity results for these materials. The room-temperature values of the conductivity of C (natural abundance diamond), SiC, AlN, and Si are 2000, 490, 320, and $160\,\mathrm{Wm^{-1}K^{-1}}$, respectively.

Table 1.1. Room-temperature values of $\mathcal{K}$ for high-thermal-conductivity nonmetallic single crystals. Units are W m^{-1} K^{-1}.

Crystal	$\mathcal{K}$	Crystal	$\mathcal{K}$	Crystal	$\mathcal{K}$
Diamond	2000	BP	360	GaN	170
BN (cubic)	1300	AlN	320	Si	160
SiC	490	BeS	300	AlP	130
BeO	370	BAs	210	GaP	100

1.5.4 Conductivity of Diamond-Structure Single Crystals

It will be interesting to discuss the temperature variation of the thermal conductivity of Si and C, two of the high-thermal-conductivity materials with diamond structure. The results for Si in Fig. 1.13, obtained theoretically using a *model effective relaxation time method* [32], are in excellent agreement with experimental results. At low temperatures, when the boundary scattering mechanism dominates, the conductivity follows the expected increase as T^3. As the temperature increases, mass-defect scattering becomes important and reduces the T^3 dependence. Three-phonon scattering sets in at finite temperatures, becoming dominant at high temperatures. As a result, the conductivity reaches a maximum at around 20 K before starting to decrease. With only three-phonon interactions included the theory predicts a linear decrease of

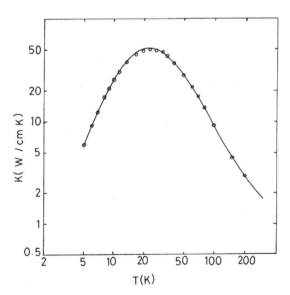

Fig. 1.13. Thermal conductivity of crystalline Si. The solid curve shows theoretically calculated results; the open circles present experimental results. Taken from [32].

conductivity above room temperature. The theory also predicts that most heat (75% or more) is conducted by transverse phonons at all temperatures.

Figure 1.14 shows measured and calculated thermal-conductivity results for single crystals of diamond with different isotopic concentrations. The theoretical calculations [33] were made using the conductivity expression in Eq. (1.44) and employing a simplified and empirically adjusted form of $\tau^{-1}(3\,ph)$. As expected, at low temperatures, when the boundary scattering dominates, the conductivity increases as T^3. As the temperature increases, isotope scattering becomes important and reduces the T^3 dependence. With a further increase in temperature, three-phonon processes become the dominant scattering mechanism. As a result, the conductivity reaches around 100 K before starting to decrease. The precise value of the conductivity maximum and its temperature variation just before reaching the maximum depend on

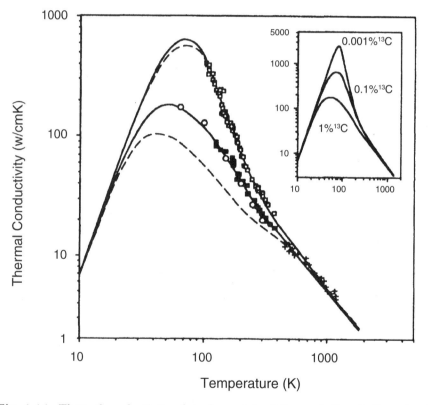

Fig. 1.14. Thermal conductivity of single crystals of diamond. Results for natural abundance diamond (1.1% ^{13}C) are: theoretical (lower solid curve) and experimental (circles and filled squares). Results for isotopically enriched sample (0.1% ^{13}C) are: theoretical (upper solid curve) and experimental (open squares). Theoretical results for some other isotopic concentrations are shown in the inset. Taken from [33] with permission.

the isotope concentration. A maximum of $41,000\,\mathrm{Wm^{-1}K^{-1}}$ is obtained for a 99.9% pure ^{12}C crystal [33]. Well above room temperature, the conductivity decreases nearly linearly with temperature rise, in accordance with the theoretical prediction.

1.6 Conductivity of Polycrystalline Solids

The usefulness of germanium-silicon alloys as high-temperature thermoelectric materials has long been realized. Similarly, high-thermal-conductivity materials such as SiC, β-Si$_3$N$_4$, and AlN are considered to be potential candidates for use in thermal management of semiconductor processing equipments. However, the high cost of producing single crystals has necessitated research on the use of powdered forms of such materials. As can be expected, the thermal conductivity of powders is significantly lower than their crystalline counterparts. The usual method of producing ceramics is by powder processing and sintering. Recent research has shown that material processing involving the use of high-purity nitride powders, hot-pressing (pressure sintering), and additives leads to a significant increase in the thermal conductivity of ceramics and thin films, with values close to those for single crystals (see Watari et al. [34] and references therein).

Investigations by Goldsmidt and Penn [35] and by Parrott [36], based on the relaxation-time theory described in this chapter, predict that sintered materials should have considerably smaller thermal conductivity than their single-crystal counterparts. The Goldsmidt-Penn-Parrott law for the decrease in conductivity is $\Delta W/W \propto \sqrt{L}$, where ΔW is the increase over the single-crystal thermal resistivity W and L is the average grain size. This law was verified in recent experimental investigations by Akimune et al. [37] for the thermal conductivity of sintered β-Si$_3$N$_4$ containing Y$_2$O$_3$-Nd$_2$O$_3$ additives and one-dimensional aligned β-Si$_3$N$_4$ whiskers. Calculations made by Kitayama et al. [38], based on a simple theoretical model for the thermal conductivity of a composite material, predict that the thermal conductivity of β-Si$_3$N$_4$ decreases quickly as the grain-boundary film thickness increases within a range of a few tenths of a nanometer and then reaches almost a constant value for larger grain sizes. Their theoretical and experimental studies demonstrate that grain growth alone cannot improve the thermal conductivity of this material. Consistent with this work, Watari et al. [39] have shown, from theoretical and experimental investigations, that for large grain-size distributions (in the μm range) the thermal conductivity of Si$_3$N$_4$ at room temperature is independent of grain size. The conductivity in such samples is found to be controlled by the internal defect structure of the grains, such as point defects and dislocations. This scenario is in agreement with an earlier suggestion made by Meddins and Parrott [40] for the possible explanation of the conductivity of sintered germanium-silicon alloys. A grain-growth-assisted percolation model has been proposed by Pezzotti [41] to explain the increase in the thermal conductivity on annealing of an AlN polycrystal doped with Y_2O_3. The model considers the concurrent effect of the growth of AlN-matrix grains and of the collapse of grain boundaries tilled by the low-thermal-conductivity Y_2O_3 phase. As a result the size of the thermally conductive AlN clusters grows and there is a reduction in the number of grain boundaries.

1.7 Conductivity of Low-Dimensional Solids

In general, a low-dimensional system lacks the perfect three-dimensional periodicity of a single crystal. Such systems can either occur naturally or be grown in a laboratory. In the context of this chapter on high-thermal-conductivity materials we will consider superlattices, graphite, graphene and carbon nanotubes as examples of low-dimensional solids.

1.7.1 Superlattices

Consider a superlattice $(A)_{N_1}(B)_{N_2}[hkl]$, made from growth of alternating N_1 layers of material A and N_2 layers of material B along a direction $[hkl]$. For simplicity, let us assume that the materials A and B are of the same crystal type and that both retain their intrinsic periodicity in each layer. Then, while the superlattice structure retains the intrinsic bulk periodicity perpendicular to $[hkl]$, the periodicity along $[hkl]$ has increased to $d = d_1 + d_2$, where d_1 and d_2 are the layer thicknesses of the materials A and B, respectively. Using the reciprocal-space language, there is a "minizone" formation along the growth direction $[hkl]$, with the superlattice Brillouin zone (SL-BZ) boundary being at π/d.

In the context of our discussion of lattice thermal conductivity, we will only discuss phonons in the bulk acoustic region, as these are the dominant heat carriers. The dispersion of such phonons can be approximated as the average of the dispersions in the bulk materials A and B. Along the superlattice growth direction $[hkl]$, however, the dispersion can be considered as the "folding" of the bulk dispersion curves, in accordance with the "minizone" formation. This effect is illustrated in Fig. 1.15 using the example of a superlattice in the form of a linear chain.

In general, as can be expected, the thermal conductivity of a superlattice along its growth direction will be lower than the weighted average value obtained from its constituent bulk materials. The difference may be contributed by three "extra" sources of phonon scattering in the superlattice: mass disorder, strain and chemical effects, and "mini-Umklapp" processes. The mass disorder due to superlattice formation can be expressed as $\Delta m/m$, where Δm is the atomic mass difference between the superlattice layer materials A and B. Superlattice strain and chemical effects arise when the materials A and B do not have the same lattice constants, and the chemical bonds between A–A, B–B, and A–B are not very similar. Additional phonon

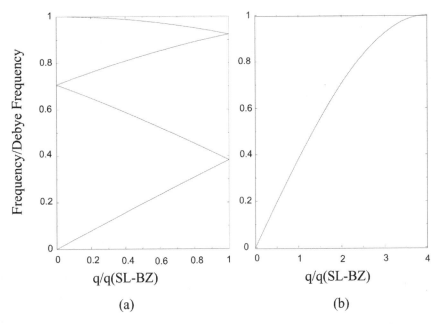

Fig. 1.15. Schematic illustration of the zone-folding effect in the phonon-dispersion curve of a monatomic linear chain of atoms: (a) the dispersion curve for a superlattice, and (b) the dispersion curve for an average bulk. q(SL-BZ) represents the superlattice Brillouin zone boundary.

anharmonic interactions in the form of "mini-Umklapp" processes will result from the formation of the mini (or superlattice) zone. This is illustrated in Fig. 1.16: if the sum of the phonon vectors $\mathbf{q}$ and $\mathbf{q}'$ lies outside the superlattice Brillouin zone (but inside the bulk zone), it can be flipped back to an equivalent point inside the superlattice zone with the help of a superlattice reciprocal lattice vector $\mathbf{G}$(SL). Theoretical work by Ren and Dow [42] predicts a significant reduction of the conductivity peak due to mini-U processes. Recent theoretical work [43] suggests that the thermal conductivity of a superlattice should show a minimum when its layer thickness is somewhat smaller than the mean free path of the phonons. Although quantitative modeling of thermal conductivity of superlattices is a rather complex task, in a general sense the preceding considerations are consistent with experimental conductivity measurements for Si/Ge [44] and GaAs/AlAs [45] superlattices.

1.7.2 Semiconductor Quantum Wells and Wires

The lattice thermal conductivity of semiconductor quantum wells and wires has been studied by adopting two distinctly different approaches: (a) the single-mode relaxation-time method, as described in Sects. 1.2.3 and 1.5.1, and (b) a molecular dynamics method [46], in which thermodynamic and

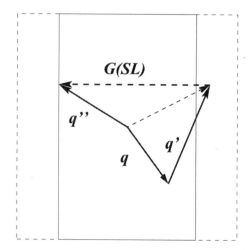

Fig. 1.16. Schematic illustration of the mini-Umklapp process due to superlattice formation. The vector $\mathbf{G}$(SL) represents a superlattice reciprocal lattice vector. For clarity, the bulk Brillouin zone is drawn by thin dashed lines.

transport properties are calculated from a numerical simulation of particle trajectories in the system based on an empirical form of interatomic potential. The single-mode relaxation-time approach, based on the solution of the Boltzmann equation, has been followed in Ref. [47] for a free-standing semiconductor quantum well and in Refs. [48] and [49] for wires.

Confined acoustic modes in a quantum well are characterized as shear (S), dilatational (D), and flexural (F) waves [50]. The S modes are similar to the transverse modes in bulk but with the displacement vector lying in the plane of the quantum well and pointing perpendicular to the direction of the in-plane wave-vector. The D and F waves can be viewed as a modification of the bulk longitudinal mode. The work carried out by Balandin and Wang [47] shows that strong modifications of the phonon dispersion and group velocities due to spatial confinement lead to a significant increase of the relaxation rates due to three-phonon scattering, impurity scattering, and boundary-scattering processes. Their numerical calculations indicate that the values of the room-temperature thermal conductivity of 155-nm-wide and 10-nm-wide Si quantum wells are approximately 45% and 13%, respectively, of the bulk value.

Zou and Balandin [49] developed a model for heat conduction in semiconductor wires. Their model is based on the solution of the phonon Boltzmann equation, taking into account (i) modification of the bulk acoustic phonon dispersion due to wire formation and (ii) the change in the nonequilibrium phonon distribution due to partially diffuse boundary scattering. Numerical calculations by Zou and Balandin for a free-standing continuum of a cylindrically symmetric nanowire show significant departure of the confined acoustic branches from the bulk longitudinal branch. This feature would obviously lead to very different phonon-relaxation rates in nanowires. In addition, the role of specular phonon-boundary scattering becomes an important consideration in calculating the thermal conductivity of a nanowire. Indeed, numerical calculations by Walkauskas et al. [48] and by Zou and Balandin [49] clearly show that the lattice thermal conductivity of nanowire structures is dramatically controlled by phonon-boundary scattering. Even in the case of purely specular boundary scattering (when deviation from the bulk conductivity due to the phonon redistribution by the boundaries vanishes), it is found that phonon confinement effects lead to deviation of the thermal conductivity from its bulk value. Numerical calculations carried out by Zou and Balandin [49] predict that for purely diffusive boundary scattering the room-temperature thermal conductivity of a cylindrical Si nanowire of diameter 20 nm would be reduced to about 9% of its bulk value.

1.7.3 Graphite, Graphene, Carbon Nanotubes, and Fullerenes

As discussed in Sect. 1.3.2, graphite, graphene, and nanotubes can be considered as quasi-two dimensional, two-dimensional (2D), and quasi-one dimensional materials, respectively. In that section we discussed the main features of lattice vibrations in these materials. When applying the single-mode relaxation-time theory of lattice thermal conductivity, we can regard a single graphene sheet as a two-dimensional phonon gas. The phonon-dispersion curves of graphite have an almost two-dimensional character, except for modes below $\nu_c \simeq 4$ THz. As discussed in Sect. 1.3.2, the phonon spectrum of a nanotube formed by rolling a single graphene sheet is continuous along its tube axis, similar to that of a graphene sheet. The phonon spectrum of a nanotube formed by rolling a few graphene sheets contains a set of such continuous spectra for each circumferential wave vector, with the lowest set corresponding to progressive breathing modes. These features of lattice vibrations along the axis of nanotubes, of graphene, and in the basal plane of graphite would allow the thermal conductivity of these three forms of carbon to be modeled by using the concept of two-dimensional phonon gas [51, 52, 53]. Here we will briefly outline the theory of thermal conductivity in two dimensions.

The single-mode relaxation-time approach, within a dispersive continuum model and a Debye-like approach, was employed more than three decades ago by Dreyfus and Maynard [54] to successfully explain the thermal conductivity of the basal plane of graphite. Here we present a simplified version of the theory within the isotropic continuum model. Following Eq. (1.23), the thermal conductivity in the basal plane of graphite can be expressed as

$$\mathcal{K}_{\text{smrt}-2D} = \frac{1}{2} \frac{\hbar^2}{N_0 \Omega k_B T^2} \sum_{\mathbf{q}s} c_s^2(\mathbf{q}) \omega^2(\mathbf{q}s) \tau_{\mathbf{q}s} \bar{n}_{\mathbf{q}s} (\bar{n}_{\mathbf{q}s} + 1), \qquad (1.46)$$

where $1/2$ represents the two-dimensional average of $\cos^2 \theta$, with θ being the angle between an applied temperature gradient and a phonon velocity. To simplify this expression we replace the sum over $\mathbf{q}$ by integration and note that the density of states of the two-dimensional system can be expressed within the isotropic continuum model as [1]

$$g(\omega) = \frac{N_0 A}{2\pi} \frac{\omega}{c^2}, \tag{1.47}$$

where A represents the area of the two-dimensional unit cell. We can further replace the Brillouin zone by a circular cylinder of radius q_m and height $q_{m\perp}$. The conductivity integral then becomes

$$\mathcal{K} = \frac{\hbar^2 q_m^4 q_{m\perp}}{4\pi^2 k_B T^2} \sum_s c_s^4 \int_{x_c}^1 dx\, x^3 \tau \bar{n}(\bar{n}+1), \tag{1.48}$$

with $x = q/q_m$ and only two phonon polarizations (longitudinal and fast transverse) considered in the basal plane so that $2/\bar{c}_{2D}^2 = 1/c_1^2 + 1/c_2^2$.

Apart from the consideration of phonon-dispersion relations and construction of a two-dimensional Debye model for integration in reciprocal space, care must also be taken to rederive phonon-scattering rates for two-dimensional systems. The boundary scattering has the same form as in the three-dimensional case, except that the phonon mean free path L is limited by the dimensions of the two-dimensional system. The mass-defect and three-phonon scattering rates will, however, show frequency dependence different from three-dimensional systems. Using the two-dimensional density of states expression in Eq. (1.47), and following the procedure for deriving Eq. (1.34), the mass-defect scattering rate of phonons in two-dimensional systems can be obtained as

$$\tau_{\mathbf{q}s}^{-1}(\text{md} - 2D) = \frac{\pi \Gamma_{\text{md}}}{\omega_m^2} \omega^3(\mathbf{q}s), \tag{1.49}$$

where $\omega_m = \bar{c}_{2D} q_m$ is the two-dimensional Debye frequency. Clearly the significant difference lies in the frequency dependence: ω^3 for two-dimensional systems against ω^4 for three-dimensional systems. Klemens and Pedraza [51] have, however, shown that the expression for the three-phonon scattering rate in two-dimensional systems is formally the same as in three-dimensional systems, except for the replacement of the Debye frequency ω_D with the Debye-like frequency ω_m.

The expression in Eq. (1.48) has been employed by Klemens [52, 53] to discuss the conductivity of graphite, graphene, and nanotubes. It should, however, be pointed out that the low-lying modes in graphite and the breathing modes propagating parallel to the nanotubes, require careful consideration. In particular, when considering the intrinsic conductivity (due only to anharmonic phonon interactions) the lower limit in the integral in Eq. (1.48) should be taken as $x_c = \omega_c/\omega_m$. For graphite $\nu_c = 4$ THz, $\nu_m = 45.9$ THz, giving $x_c = 0.09$. For graphene $\nu_c = 0$, giving $x_c = 0$. For the (10,10) carbon nanotube the breathing mode frequency is $\nu_c = 5$ THz. But these modes can anharmonically decay into smaller modes. Klemens [53] thus considered $\nu_c = 3$ THz and $\nu_m = 46$ THz, giving $x_c = 0.07$. It is also useful to point out that the conductance κ, defined as $\mathbf{Q} = -\kappa w \, \nabla T$, where w is the width of the sheet normal to the temperature gradient, is a better measure of two-dimensional heat transport.

Applying these considerations, Klemens [51, 52, 53] has predicted a logarithmic divergence of the intrinsic conductivity (i.e., conductivity due only to three-phonon processes) of two-dimensional systems. This, he explains, happens for two reasons: (i) interaction of low frequencies outside the two-dimensional phonon spectrum with the low-frequency modes and (ii) limitation of the phonon mean free path by the external dimensions of the two-dimensional system. Factor (i) operates in graphite; reason (ii) is the case in the single graphene sheet. In carbon nanotubes either factor (i) or (ii) can operate. Employing simple expressions for three-phonon scattering rates, Klemens has estimated that the intrinsic conductivity of graphite at room temperature would be $1.9 \times 10^3 \, \mathrm{Wm^{-1}K^{-1}}$. For a graphene layer of dimension 1 cm, the room-temperature values are $\mathcal{K} = 5.5 \times 10^3 \, \mathrm{W \, m^{-1} \, K^{-1}}$ and $\kappa = 18 \times 10^{-7} \, \mathrm{WK^{-1}}$. He has further estimated that the conductivity of a long carbon nanotube would be slightly higher than the conductivity in the basal plane of graphite: $\mathcal{K}(\mathrm{nano}) \simeq 1.18\mathcal{K}(\mathrm{graphite})$.

It is interesting to discuss the low-temperature behavior of the thermal conductivity of carbon nanotubes. It is worth pointing out that, depending on their helical structures, carbon nanotubes can be metallic or semiconducting, but here we are only interested in discussing their lattice thermal conductivity contributed by phonons. The features of phonon-dispersion curves responsible for low-temperature specific heat and the conductivity along the tube axis depend strongly on the tube diameter. Theoretical work by Benedict et al. [55] has predicted that at low temperatures (i.e., much below Debye temperature) the specific heat exhibits a quasi-one-dimensional behavior: for small tube radius R, and lower temperatures, $C_v^{\mathrm{sp}} \propto T$; otherwise, $C_v^{\mathrm{sp}} \propto T^2$. This behavior is schematically illustrated in Fig. 1.17. The heat capacity of

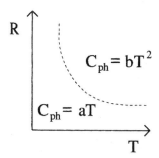

Fig. 1.17. Schematic illustration of the temperature variation of the specific heat of carbon nanotubes. At low temperatures (much lower than Debye temperature), $C_v^{\mathrm{sp}} \propto T$ for small tube radius and lower temperatures, and $C_v^{\mathrm{sp}} \propto T^2$ otherwise. Taken from Benedict et al. [55] with permission.

a narrow nanotube of radius R (so that $T \ll \hbar\bar{c}/k_\mathrm{B}R$) of a single acoustic phonon polarization is [55]

$$C_\mathrm{v}(\text{nanotube}) = \frac{3Lk_\mathrm{B}^2 T}{\pi\hbar v A} 3.292, \tag{1.50}$$

where A is the cross-sectional area of the tube, L is the tube length, and $\bar{c}$ is the average acoustic speed. The specific heat capacity of the nanotube (specific heat per unit length) is then $C_\mathrm{v}^\mathrm{sp} = C_\mathrm{v}/L$. Assuming the phonon-relaxation time is limited by the energy-independent boundary length l_B, Hone et al. [56] obtained the following expression for the low-temperature thermal conductivity of nanotubes:

$$\mathcal{K} = \frac{3.292 k_\mathrm{B}^2 l_b T}{\pi\hbar A}. \tag{1.51}$$

This result shows excellent fit to the experimental data obtained by Hone et al. [56] for the conductivity of single-walled nanotubes below 30 K.

Detailed nonequilibrium molecular dynamics simulations of the thermal conductivity of carbon nanotubes have been performed by Berber et al. [57]. The results for an isolated (10,10) nanotube over a large temperature range are shown in Fig. 1.18. At low temperatures the results are in agreement with the measurements made by Hone et al. [56]. The maximum conductivity occurs at 100 K, with an unusually high value of $37,000\,\mathrm{Wm^{-1}K^{-1}}$. The maximum conductivity of the nanotube is close to the maximum value of $41,000\,\mathrm{Wm^{-1}K^{-1}}$ for a 99.9% pure ^{12}C crystal at 104 K [33]. The room-temperature thermal conductivity of the nanotube is predicted to be $6600\,\mathrm{W\,m^{-1}\,K^{-1}}$, comparable to that of a hypothetical graphene monolayer and exceeding the conductivity value of $3320\,\mathrm{Wm^{-1}K^{-1}}$ for 99.9% ^{12}C, isotopically enriched diamond [58]. It is believed that the high conductivity values of the nanotubes are associated with strong sp^2 bonding configuration and large-phonon mean free paths in these systems.

Finally, we briefly discuss the thermal conductivity of single-crystal fullerenes. The main contribution to the thermal conductivity of fullerenes

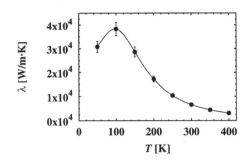

Fig. 1.18. Theoretical prediction of the temperature variation of the lattice thermal conductivity of the (10,10) carbon nanotube. Taken from Berber et al. [57] with permission.

Table 1.2. Room-temperature values of the thermal conductivity $\mathcal{K}$ of different solid forms of carbon. The values for graphene and nanotube are theoretical estimates. References are cited in the main text. Units are $\mathrm{Wm^{-1}K^{-1}}$.

Material	$\mathcal{K}$
Diamond (natural abundance)	2000
Graphite (basal plane)	2000
Graphene	3320
Carbon nanotube	6600
C_{60}	0.4

is from the lattice, with free carriers contributing less than 10% [59]. The magnitude and temperature variation of the lattice thermal conductivity of C_{60} are similar to those observed for a glassy solid and show particular sensitivity to orientational disorder [60, 61]. The orientational disorder is evidenced by the observed time dependence of the phonon mean free path [61]. A simple phenomenological model, involving a thermally activated jumping motion between two nearly degenerate orientations, has been found to describe the temperature and time dependence of the thermal conductivity [61]. The room-temperature thermal conductivity of single-crystal C_{60} is approximately $0.4\,\mathrm{Wm^{-1}K^{-1}}$ [61]. Table 1.2 summarizes the room-temperature values of the thermal conductivity of different solid forms of carbon.

1.8 Summary

In this chapter I have discussed the fundamental aspects of the theroy of lattice thermal conductivity of nonmetallic solids in crystalline (3D), polycrystalline, and low-dimensional (2D and 1D) forms. The role of phonon-dispersion relations and anharmonic phonon interactions, in particular three-phonon processes, in developing the theory of thermal conductivity has been emphasized. Explicit expressions of phonon-relaxation times and the conductivity within the single-mode relaxation-time approach have been produced. The theory presented here has only one possible adjustable parameter, Grüneisen's anharmonic coefficient γ. The simplified intrinsic conductivity expression, within the high-temperature approximation, has been used to derive a set of rules for choosing high-thermal-conductivity materials. The theory has been applied to discuss the conductivity results for solids in bulk, powder, and low-dimensional forms. In particular, conductivity results of quantum wells, quantum wires, and the various solid forms of carbon—diamond, graphite, graphene, nanotubes, and C_{60} fullerenes—have been presented and discussed.

References

[1] G. P. Srivastava, *The Physics of Phonons* (Adam Hilger, Bristol, 1990).

[2] R. A. Guyer and J. A. Krumhansl, *Phys. Rev.* **148**, 766 (1966).

[3] R. Peierls, *Ann. Inst. H. Poincaré* **5**, 177 (1935).

[4] P. G. Klemens, *Proc. R. Soc. A* **208**, 108 (1951).

[5] G. Leibfried and E. Schlömann, *Nachr. Akad. Wiss. Göttingen Math. Phys. Kl II(a)* **4**, 71 (1954).

[6] J. M. Ziman, *Electrons and Phonons* (Clarendon, Oxford, 1960).

[7] R. A. H. Hamilton and J. E. Parrott, *Phys. Rev.* **178**, 1284 (1969).

[8] A. M. Arthurs *Complementary Variational Principles* (Oxford: Clarendon, 1970).

[9] D. Benin, *Phys. Rev. B* **1**, 2777 (1970).

[10] G. P. Srivastava, *J. Phys. C: Solid State Phys.* **9**, 3037 (1976).

[11] J. E. Parrott and A. D. Stukes, *Thermal Conductivity of Solids* (Pion Ltd., London, 1975).

[12] P. Debye, *Vortraege über die kinetische Theorie der Materie und der Elektrizitaet* (Teubner, Berlin, 1914), p. 19

[13] P. G. Klemens, *Solid State Physics* vol 7, (ed. F. Seitz and D. Turnbull, Academic, New York, 1958) p. 1.

[14] J. Callaway, *Phys. Rev.* **113**, 1046 (1959).

[15] S. Simons, *J. Phys. C: Solid State Phys.* **8**, 1147 (1975).

[16] G. P. Srivastava, *Phil. Mag.* **34**, 795 (1976).

[17] G. P. Srivastava, *Theoretical Modelling of Semiconductor Surfaces* (Singapore: World Scientific, 1999).

[18] W. Weber, *Phys. Rev. B* **15**, 4789 (1977).

[19] H. M. Tütüncü and G. P. Srivastava, *Phys. Rev. B* **62**, 5028 (2000).

[20] R. Nicklow, N. Wakabayashi, and H. G. Smith, *Phys. Rev. B* **5**, 4951 (1972).

[21] R. Saito, G. Dresselhaus, and M. S. Dresselhaus, *Physical Properties of Carbon Nanotubes* (Imperial College Press, London, 1998).

[22] J. Yu, R. K. Kalia, and P. Vashistha, *J. Chem. Phys.* **103**, 6697 (1995).

[23] P. G. Klemens, *Proc. Phys. Soc. A* **68**, 1113 (1955).

[24] Y. P. Joshi, *Phys. Stat. Solidi (b)* **95**, 627 (1979).

[25] D. J. Ecsedy and P. G. Klemens, *Phys. Rev. B* **15**, 5957 (1977).

[26] C. Herring, *Phys. Rev.* **95**, 954 (1954).

[27] G. L. Guthrie, *Phys. Rev.* **152**, 801 (1966).

[28] G. P. Srivastava, *Pramana* **3**, 209 (1974).

[29] G. P. Srivastava, *Pramana* **6**, 1 (1976).

[30] J. E. Parrott, *Proc. Phys. Soc.* **81**, 726 (1963).

[31] M. Roufosse and P. G. Klemens, *Phys. Rev. B* **7**, 5379 (1973).

[32] G. P. Srivastava, *J. Phys. Chem. Solids* **41**, 357 (1980).

[33] L. Wei, P. K. Kuo, R. L. Thomas, T. R. Anthony, and W. F. Banholzer, *Phys. Rev. Lett.* **70**, 3764 (1993).

[34] K. Watari, K. Hirao, M. E. Brito, M. Toriyama, and S. Kanzaki, *J. Mater. Res.* **14**, 1538 (1999).

[35] H. J. Goldsmidt and A. W. Penn, *Phys. Lett.* **27A**, 523 (1968).

[36] J. E. Parrott, *J. Phys. C (Solid St. Phys.)* **2**, 147 (1969).

[37] Y. Akimune, F. Munakata, K. Matsuo, Y. Okamoto, N. Hirokasi, and C. Satoh, *J. Ceram. Soc. Jpn.* **107**, 1180 (1999).

[38] M. Kitayama, K. Hirao, M. Toriyama, and S. Kanzaki, *J. Am. Ceram. Soc.* **82**, 3105 (1999).

[39] K. Watari, K. Hirao, M. Toriyama, and K. Ishizaki, *J. Am. Ceram. Soc.* **82**, 777 (1999).

[40] H. R. Meddins and J. E. Parrott, *J. Phys. C (Solid St. Phys.)* **9**, 1263 (1976).

[41] G. Pezzoti, *J. Ceram. Soc. Jpn.* **107**, 944 (1999).

[42] S. Y. Ren and J. D. Dow, *Phys. Rev. B* **25**, 3750 (1982).

[43] M. V. Simkin and G. D. Mahan, *Phys. Rev. Lett.* **84**, 927 (2000).

[44] S.-M. Lee, D. G. Cahill, and R. Venkatasubramanium, *Appl. Phys. Lett.* **70**, 2957 (1997).

[45] W. S. Capinski, H. J. Maris, T. Ruf, M. Cardona, K. Ploog, and D. S. Katzer, *Phys. Rev. B* **59**, 8105 (1999).

[46] S. G. Volz and G. Chen, *Appl. Phys. Lett.* **75**, 2056 (1999).

[47] A. Balandin and K. L. Wang, *Phys. Rev. B* **58**, 1544 (1998).

[48] S. G. Walkauskas, D. A. Broido, K. Kempa, and T. L. Reinecke, *J. Appl. Phys.* **85**, 2579 (1999).

[49] J. Zou and A. Balandin, *J. Appl. Phys.* **89**, 2932 (2001).

[50] N. Bannov, V. Aristov, V. Mitin, and M. A. Stroscio, *Phys. Rev. B* **51**, 9930 (1995).

[51] P. G. Klemens and D. F. Pedraza, *Carbon* **32**, 735 (1994).

[52] P. G. Klemens, *J. Wide Bandgap Materials* **7**, 332 (2000).

[53] P. G. Klemens, *Proc. 26th Int. Thermal Cond. Conf.* (6–8 Aug. 2001, Cambridge, Mass., USA.).

[54] P. B. Dreyfus and R. R. Maynard, *J. Physique (France)* **28**, 955 (1967).

[55] L. X. Benedict, S. G. Louie, and M. L. Cohen, *Solid St. Commun.* **100**, 177 (1996).

[56] J. Hone, M. Whitney, C. Piskoti, and A. Zettl, *Phys. Rev. B* **59**, R2514 (1999).

[57] S. Berber, Y. K. Kwon, and D. Tomanek, *Phys. Rev. Lett.* **84**, 4613 (2000).

[58] T. R. Anthony, W. F. Banholzer, J. F. Fleischer, L. Wei, P. K. Kuo, R. L. Thomas, and R. W. Pryor, *Phys. Rev. B* **42**, 1104 (1990).

[59] K. Biljaković, A. Smontara, D. Starešinić, D. Pajić, M. E. Kozlov, M. Hirabayashi, M. Tokumoto, and M. Ihara, *J. Phys.: Condens. Matter* **8**, L27 (1996).

[60] K. Biljaković, M. Kozlov, D. Starešinić, and M. Saint-Paul, *J. Phys.: Condens. Matter* **14**, 6403 (2002).

[61] R. C. Yu, N. Tea, M. B. Salamon, D. Lorents, and R. Malhotra, *Phys. Rev. Lett.* **68**, 2050 (1992).

2

High Lattice Thermal Conductivity Solids

Donald T. Morelli and Glen A. Slack

The lattice thermal conductivity κ of various classes of crystalline solids is reviewed, with emphasis on materials with $\kappa > 0.5\,\mathrm{Wcm^{-1}K^{-1}}$. A simple model for the magnitude of the lattice thermal conductivity at temperatures near the Debye temperature is presented and compared to experimental data on rocksalt, zincblende, diamond, and wurtzite structure compounds, graphite, silicon nitride and related materials, and icosahedral boron compounds. The thermal conductivity of wide-band-gap Group IV and Group III–V semiconductors is discussed, and the enhancement of lattice thermal conductivity by isotopic enrichment is considered.

2.1 Introduction: The Importance of Thermal Conductivity

A solid's thermal conductivity is one of its most fundamental and important physical parameters. Its manipulation and control have impacted an enormous variety of technical applications, including thermal management of mechanical, electrical, chemical, and nuclear systems; thermal barriers and thermal insulation materials; more efficient thermoelectric materials; and sensors and transducers. On a more fundamental level, the study of the underlying physics of the heat-conduction process has provided a deep and detailed understanding of the nature of lattice vibrations in solids. In this review we focus on solid electrically insulating materials with high lattice thermal conductivity. By lattice thermal conductivity we mean heat conduction via vibrations of the lattice ions in a solid. Our goal is to first provide a simple physical picture for lattice heat conduction in solids and to then compare this model with experimental data on the thermal conductivity of several classes of crystal structures and types of materials. The review is similar in spirit to that of Slack [1] but incorporates and discusses data and experimental results that have been obtained since that review. The present work is mainly concerned with the intrinsic lattice thermal conductivity of solids. Klemens [2] has reviewed the influence of various types of defects and impurities on the lattice thermal conductivity. The classic monograph of Berman [3] discusses all aspects of the thermal conductivity of solids, including metals, polymers, and amorphous materials. A more recent update on materials advances in the area of high thermal conductivity has also recently appeared in the literature [4].

The fact that certain materials that are good electrical insulators can possess high thermal conductivity is frequently met with surprise and puzzlement by the casual observer. This is easily understood, however, when one realizes that whereas electrical current in a material is carried solely by charge carriers, heat may be transported by both charge carriers and vibrations of the lattice ions. In a good metal like copper, the electron density is large, and nearly all of the heat conduction occurs via charge carrier transport. This electronic thermal conductivity masks the lattice thermal conductivity, which is present but small relative to the electronic term. In a material where there are no free electrons to carry heat, the lattice thermal conductivity is the only mode of heat transport available. Within the family of electrically insulating materials, the magnitude of the lattice thermal conductivity, κ, can vary over an extremely wide range. For instance, diamond has a thermal conductivity at room temperature of $30\,\mathrm{Wcm^{-1}K^{-1}}$, much higher than that of any material, including the best metals. On the other hand, some polymeric materials and amorphous electrically insulating solids have thermal conductivity at room temperature as low as $0.001\,\mathrm{Wcm^{-1}K^{-1}}$. We want to understand why certain materials can possess high lattice thermal conductivity and what physical mechanisms serve to provide a limit to the lattice thermal conductivity of solids.

The review is organized as follows. In Sect. 2.2 we will introduce simple models of lattice heat conduction that can be used to predict the magnitude and temperature dependence of the thermal conductivity. In Sect. 2.3 we consider some specific classes of materials that possess high thermal conductivity and compare experimental results with the predictions of this model. Sect. 2.4 takes a closer look at lattice heat conduction in several technologically important wide-band-gap semiconductors. In Sect. 2.5 we discuss how the isotope effect may be used to increase the lattice thermal conductivity of some materials. Finally, Sect. 2.6 provides a summary and suggests some future directions of research on high-thermal-conductivity solids.

Of course we must first define what we mean by "high" thermal conductivity. As mentioned previously, the lattice thermal conductivity of solids near ambient temperature can span an enormously wide range. "High" thermal conductivity is thus a relative term; for instance, a polymer with a thermal conductivity of $0.03\,\mathrm{Wcm^{-1}K^{-1}}$ would, for this class of solids, have a "high" thermal conductivity. On the other hand, such a value of thermal conductivity for an inorganic crystalline semiconductor (the thermoelectric material PbTe, for example) would be considered very "low". Frequently in the literature a value of thermal conductivity in excess of $1\,\mathrm{Wcm^{-1}K^{-1}}$ has been chosen, rather arbitrarily, as the lower limit for a high-thermal-conductivity solid. Because the main driver in the search for high-thermal-conductivity solids is for thermal management of electronics systems, a more suitable metric may be how the thermal conductivity compares to traditional materials used in these types of applications. By far the most widely used material for thermal management in high-volume applications is crystalline alumina, with a thermal conductivity on the order of $0.5\,\mathrm{Wcm^{-1}K^{-1}}$. We will thus set our lower limit for "high" thermal conductivity at $0.5\,\mathrm{Wcm^{-1}K^{-1}}$. As we shall see, even with this more relaxed criterion, the family of high-thermal-conductivity electrical insulators is still rather small.

2.2 Simple Model of the Magnitude of Lattice Heat Conduction in Solids

2.2.1 Normal Modes of Vibrations of a Lattice

The concepts central to an understanding of the lattice thermal conductivity of a solid are captured in the simple model of a linear chain of atoms of mass M held together by springs of force constant k. If the rest of the atoms are a distance a apart, the relation between the frequency ω and wavenumber q of a wave along the chain is given by

$$\omega(q) = 2\sqrt{\frac{k}{M}}|\sin(qa/2)|. \tag{2.1}$$

This relationship between the frequency and wavenumber of a wave is termed the dispersion curve and is illustrated in Fig. 2.1a for wavenumber ranging between $-\pi/a$ and $+\pi/a$, which represents the first Brillouin zone for the one-dimensional chain in reciprocal space. An essential feature of the relationship between frequency and wavenumber that distinguishes the present case from that of a continuum elastic wave is the bending over, or "dispersion," of the curve near the edge of the Brillouin zone. Because the group velocity of the wave is given by $v = d\omega/dq$, near these extrema the velocity of the wave tends to zero.

In a linear chain of atoms with two different types of masses, M_1 and M_2, alternating along the length of the chain, there are two solutions to the wave equation, and the resulting $\omega - q$ relations are termed the two branches of the dispersion relation. These are shown in Fig. 2.1(b). The lower branch, called the acoustic branch because the linear relationship $\omega = vq$ for low frequency is similar to that for a sound wave, is the same as that for the case of a chain of atoms of a single type, shown in Fig. 2.1(a). This branch corresponds to two neighboring atoms moving in phase with one another. The upper branch,

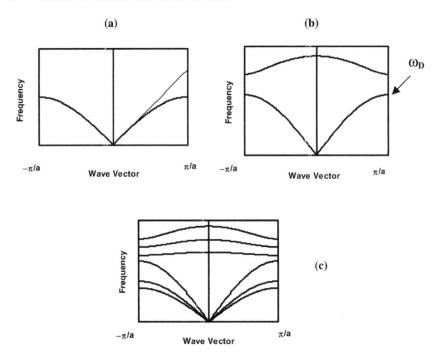

Fig. 2.1. Models of phonon-dispersion curves for solids: (a) one-dimensional case for single-atom type spaced by distance a, fine line represents continuum case; (b) one-dimensional case for two atoms with differing masses, showing the occurrence of both an acoustic (lower curve) and an optic (upper curve) branch; ω_D, where the acoustic branch meets the zone edge, is the Debye frequency for the acoustic phonons; (c) three-dimensional lattice with two different atom masses.

called the optic branch, corresponds to the case where two neighboring atoms are moving out of phase with one another; for low frequencies this branch is characterized by a vanishing group velocity. Because the group velocity of the optic branch is small, these modes generally do not participate in the heat transport process, and most of the energy transport along the chain occurs via the acoustic branch. This is a basic assumption that we use throughout this review. There are instances, however, especially at high temperatures, when this may not be true; these are touched on in Sect. 2.3. Possible heat conduction by optic phonons is considered in more detail in the review by Slack [1] and has been treated for the specific case of alkali halide compounds by Pettersson [5]. Additionally, while the optic branch is generally ineffective in transporting heat, these modes may "interact" with the heat-carrying acoustic vibrations and thus can be important in determining the magnitude of the thermal conductivity.

Of course, an actual crystal is not a linear chain of atoms but a three-dimensional lattice. In this case, if all the atoms of the lattice are of the same mass, there are three acoustic branches representing the three polarization modes (one longitudinal and two transverse) of the crystal. If there is more than one type of atom per unit cell, the dispersion relation again will contain optic modes. As in the one-dimensional case, these modes are typified by high frequency and low group velocity. The dispersion curve for a three-dimensional lattice containing two different types of atoms is shown in Fig. 2.1(c). For the more general case of N types of atoms, there will be three acoustic branches and $3(N-1)$ optic branches.

Of fundamental importance in the heat transport in a lattice is the concept of the Debye frequency, ω_D, which is defined here as the maximum vibrational frequency of a given mode in a crystal. For acoustic modes, this corresponds to the frequency at the zone boundary as indicated in Fig. 2.1(b) for the one-dimensional chain. One can define a Debye temperature θ_a for an acoustic phonon branch as:

$$\theta_a = \frac{\hbar \omega_D}{k_B},\qquad (2.2)$$

where $\hbar$ is the Planck constant and k_B is the Boltzmann constant. For the three-dimensional case each acoustic branch will have a Debye temperature given by Eq. (2.2) with its appropriate value of ω_D.

An alternative method of calculating θ_a is by integrating the acoustic portion of the phonon density of states $g(\omega)$ over the Brillouin zone according to [6]:

$$\theta_a^2 = \frac{5\hbar^2}{3k_B^2} \frac{\int\limits_0^{\omega_D} \omega^2 g(\omega) d\omega}{\int\limits_0^{\omega_D} g(\omega) d\omega}. \qquad (2.3)$$

For $g(\omega) \sim \omega^2$ these two definitions are equivalent. The Debye temperature can be thought of as the temperature above which all vibrational modes in a crystal are excited.

Clearly, given the dispersion relations of real crystals, the Debye temperature for heat transport is determined by where the acoustic branches of the vibrational spectrum meet the Brillouin zone edge; this will be different from the Debye temperature determined by other means, such as calculation from the elastic constants or from the low-temperature specific heat. The importance of this point has been discussed in detail by Slack [1] and will become clearer as our discussion continues.

Although the preceding discussion is useful for describing the frequency–wave vector relationships for phonons, it is insufficient for a discussion of the thermal conductivity. This is because in the presence of only harmonic interactions there is no means of interaction between different phonons, and in such a situation the mean free path for lattice vibrations would be infinite. Only by considering the higher-order anharmonic terms in the ionic interaction energy can we account for finite thermal conductivity. These higher-order terms are characterized by the Grüneisen constant γ. This "constant" is defined as the rate of change of the vibrational frequency of a given mode with volume:

$$\gamma = -\frac{d \ln \omega_i}{d \ln V} \tag{2.4}$$

and is therefore not a constant but a function of q. Again, different vibrational modes will have different values of γ. Because γ is a measure of the departure of a crystal from harmonicity, we expect that any model of the thermal conductivity will include this parameter as well. What we really want are the γ-values for the acoustic modes at temperatures on the order of the Debye temperature, which unfortunately are unavailable in most circumstances. The γ-values determined from thermal expansion data, for instance, average over all phonon branches, including the optic branches. Thus there is a great deal of uncertainty in the choice of this parameter for many of the solids we are considering in this review. Recently some lattice dynamical calculations have become available that provide mode Grüneisen parameters, and we will use these in estimating γ-values when appropriate. In other cases γ will be estimated from thermal expansion data.

2.2.2 Normal and Umklapp Phonon-Scattering Processes

Even in a perfect crystal there are interactions of phonons among themselves that tend to restore the phonon distribution to equilibrium. The interactions that give rise to thermal resistance involve powers higher than quadratic in the perturbation Hamiltonian describing the potential energy of a displaced ion in the lattice. Terms that are cubic in the displacement can be thought of as arising from three-phonon interactions, while those that are quartic arise from interactions among four phonons. Let us for simplicity consider only the cubic anharmonic term involving modes (ω_1, q_1) and (ω_2, q_2) interacting and resulting in mode (ω_3, q_3). The transition probability for the three-phonon process giving rise to this term is nonzero only if:

$$\omega_1 + \omega_2 = \omega_3 \quad \text{and} \quad q_1 + q_2 = q_3 + K, \tag{2.5}$$

where K is equal to zero for a so-called normal phonon process and equal to a reciprocal lattice vector for a so-called Umklapp process. This latter process, from the German phrase "to flip over," represents a situation in which the net phonon flux is reversed in direction. It can be shown that only Umklapp processes give rise to thermal resistance, and as a first approximation one can ignore the existence of normal processes in determining the thermal conductivity.

2.2.3 Relaxation-Time Approximation

In the *relaxation-time approximation* [3], it is assumed that the phonon distribution is restored to the equilibrium distribution at a rate proportional to the departure from equilibrium. By assuming a linear dispersion relation, the thermal conductivity can be expressed as:

$$\kappa_L = \frac{k_B}{2\pi^2 v} \left(\frac{k_B T}{\hbar}\right)^3 \int_0^{\theta_D/T} \frac{x^4 e^x}{\tau_C^{-1}(e^x - 1)^2} dx, \tag{2.6}$$

where $x = \hbar\omega/k_B T$ is dimensionless, ω is the phonon frequency, k_B is the Boltzmann constant, $\hbar$ is the Planck constant, θ_D is the Debye temperature, v is the velocity of sound, and τ_C is the total phonon-scattering relaxation time. The various processes that scatter phonons are assumed to be independent of one another and to be described by individual scattering rates τ_i^{-1} such that:

$$\tau_c^{-1} = \sum_i \tau_i^{-1}. \tag{2.7}$$

In general, the various scattering processes i will depend on both temperature and phonon frequency. In addition to the intrinsic Umklapp scattering process, a wide variety of other types of phonon-scattering mechanisms, including boundary scattering, point defect scattering, dislocation scattering, and magnetic scattering, to name just a few, have been considered in the literature; these are discussed in more detail in earlier reviews. Some of these scattering processes will be considered in more detail later.

2.2.4 Callaway Model

While it is indeed true that normal processes themselves do not give rise to thermal resistance, it is incorrect to assume that they do not influence the thermal conductivity, because they are capable of redistributing momentum and energy among phonons that are more likely to undergo a resistive scattering process. The most widely accepted model describing this process is that of Callaway [7]. In the Callaway model, the thermal conductivity is composed of two terms:

$$\kappa = \kappa_1 + \kappa_2$$

with

$$\kappa_1 = \frac{1}{3}CT^3 \int_0^{\theta/T} \frac{\tau_c(x)x^4 e^x}{(e^x - 1)^2} dx$$

and

$$\kappa_2 = \frac{1}{3}CT^3 \frac{\left[\int_0^{\theta/T} \frac{\tau_c(x)x^4 e^x}{\tau_N(x)(e^x - 1)^2} dx\right]^2}{\int_0^{\theta_L/T} \frac{\tau_c(x)x^4 e^x}{\tau_N(x)\tau_R(x)(e^x - 1)^2} dx}. \tag{2.8}$$

In these expressions, τ_R represents the scattering time due to resistive processes, τ_N the scattering time due to normal phonon processes, and $\tau_c^{-1} = \tau_R^{-1} + \tau_N^{-1}$ represents the combined scattering rate. We shall see that in some circumstances an adequate description of the thermal conductivity can be obtained using (2.6) while in others it is necessary to take into account normal phonon-scattering processes.

2.2.5 Thermal Conductivity Near the Debye Temperature

We see that a comprehensive model for the lattice thermal conductivity of a solid requires not only knowledge of the phonon spectrum and Grüneisen parameters, but also an understanding of various types of phonon-scattering rates and their temperature and frequency dependencies. Now we will concern ourselves only with an understanding of the intrinsic thermal conductivity of a solid in a temperature range where only interactions among the phonons themselves via anharmonic Umklapp processes are important. Various early estimates of the lattice thermal conductivity of a solid in this regime have been discussed by Slack [1] and Berman [3], and can be considered for our purposes as approximate expressions for the thermal conductivity at temperatures not too far removed from the Debye temperature of the solid. These estimates all take the form

$$\kappa = A \cdot \frac{M_a \theta_a^3 \delta}{\gamma^2 T}, \tag{2.9}$$

where M_a is the atomic mass of the atom, δ^3 is the volume per atom, and A is a constant. Leibfried and Schlömann [8] give the constant as $A = 5.72 \times 10^{-8}$ for δ in Angstroms and M_a in atomic mass units. Julian [9] pointed out an error in their calculation and determined the following value for A:

$$A = \frac{2.43 \cdot 10^{-8}}{1 - 0.514/\gamma + 0.228/\gamma^2}. \tag{2.10}$$

Slack [1] put $\gamma \approx 2$ in this expression and used $A = 3.04 \times 10^{-8}$. The γ-dependence of A is slight and we will allow this parameter to assume its value appropriate to the value of γ used to calculate the thermal conductivity.

2.2.6 Extension to More Complex Crystal Structures and Criteria for High Thermal Conductivity

Equation (2.9) is valid for structures containing only one atom per primitive unit cell. Using a simple counting scheme, Slack [1] extended the model to crystals with n atoms per unit cell:

$$\kappa = A \cdot \frac{\overline{M}_a \theta_a^3 \delta n^{1/3}}{\gamma^2 T}. \tag{2.11}$$

By using the Debye temperature appropriate for the acoustic modes only, this equation is a quantitative statement of our basic assumption that the optic modes in crystals with $n > 1$ do not contribute to the heat transport process.

In many circumstances, especially in considering new materials and crystal structures, the phonon-dispersion relations used to calculate θ_a are not available either experimentally or theoretically. In these cases, the acoustic-mode Debye temperature can be determined from the "traditional" definition of the Debye temperature θ (namely that determined from the elastic constants or specific heat) by using [10]

$$\theta_a = \theta n^{-1/3}. \tag{2.12}$$

With increasing n, the size of the unit cell (that is, the lattice constant a) in real space increases. This means that the Brillouin zone boundary (see Fig. 2.1) moves inward, thus cutting off phonon frequencies at smaller values as n increases. The "traditional" Debye temperature θ depends on the atomic mass and the bond strength but is independent of n. Thus Eq. (2.11) can be rewritten to display the explicit n-dependence of the thermal conductivity as:

$$\kappa = A \cdot \frac{\overline{M} \theta^3 \delta}{\gamma^2 T n^{2/3}}. \tag{2.13}$$

On the basis of Eq. (2.13) we may now list the necessary criteria for an electrically insulating solid to possess high thermal conductivity:

- high Debye temperature,
- small Grüneisen parameter, and
- small n (simple crystal structure).

2.3 Materials with High Lattice Thermal Conductivity

2.3.1 Rocksalt, Diamond, and Zincblende Crystal Structures

We can test the validity of this simple model for thermal conductivity by comparing it to experimental data. Let us begin by considering classes of solids with common values of n. The only nonmetallic crystals with $n = 1$ are the rare gas crystals, which crystallize in the simple cubic structure. These crystals, however, all have Debye temperatures less than $100\,\text{K}$, and as a result, have $\kappa < 0.5\,\text{Wcm}^{-1}\text{K}^{-1}$ and will not be considered further.

The families of crystals with $n = 2$ include the rocksalt, diamond, and zincblende structure compounds. The main members of these three families that we will consider here are shown in Tables 2.1 and 2.2, respectively, along

Table 2.1. Calculated and experimental room-temperature thermal conductivity of several rocksalt ($n = 2$) compounds. θ_a = high-temperature Debye temperature of the acoustic phonon branch; γ = high-temperature Grüneisen constant; δ^3 = volume per atom; M = average atomic mass; κ_{calc} = calculated thermal conductivity from equation (2.13); κ_{exp} = measured thermal conductivity.

Compound	θ_a (K)	γ	δ (Å)	M (amu)	κ_{calc} (Wcm^{-1}K^{-1})	κ_{exp} (Wcm^{-1}K^{-1})
LiH	615	1.28	2.04	3.97	0.159	0.15
LiF	500	1.5	2.00	12.97	0.194	0.176
NaF	395	1.5	2.31	21.00	0.179	0.184
NaCl	220	1.56	2.81	29.22	0.048	0.071
NaBr	150	1.5	2.98	51.45	0.031	0.028
NaI	100	1.56	3.23	74.95	0.013	0.018
KF	235	1.52	2.66	2.05	0.058	
KCl	172	1.45	3.14	37.27	0.038	0.071
KBr	117	1.45	3.30	59.50	0.020	0.034
KI	87	1.45	3.52	68.00	0.010	0.026
RbCl	124	1.45	3.27	60.46	0.024	0.028
RbBr	105	1.45	3.42	82.69	0.021	0.038
RbI	84	1.41	3.66	106.10	0.015	0.023
MgO	600	1.44	2.11	20.00	0.596	0.6
CaO	450	1.57	2.4	28.04	0.332	0.27
SrO	270	1.52	2.57	51.81	0.152	0.12
BaO	183	1.5	2.7	76.66	0.076	0.023
PbS	115	2	2.97	119.60	0.017	0.029
PbSe	100	1.5	3.06	143.08	0.035	0.020
PbTe	105	1.45	3.23	167.4	0.040	0.025

with the parameters needed to calculate their thermal conductivities using Eq. (2.13).

Let us consider the rocksalt compounds first; see Table 2.1. Here the Debye temperatures for acoustic phonons have been determined either from Eq. (2.2) or (2.3); in cases where both the phonon density of states and the phonon-dispersion relations are available, the calculated Debye temperatures using these two methods differ by less than 10 percent. For the Grüneisen parameters we use the data collected by Slack [1]. It should be noted that there is remarkably little variation in the γs for these rocksalts, with the majority of them lying in the range 1.5–1.9.

Figure 2.2 is a plot of the measured thermal conductivity at room temperature as a function of the calculated thermal conductivity. We see that, with data spanning a range of two orders of magnitude, Eq. (2.13) actually gives a very good description of the thermal conductivity of the rocksalt compounds. The tendency for the measured thermal-conductivity values to exceed the calculated ones has been attributed to a contribution from optic phonons. According to the criterion introduced earlier, only one rocksalt

Table 2.2. Calculated and experimental room-temperature thermal conductivity of several zincblende and diamond structure $(n = 2)$ compounds. $\theta_a =$ high-temperature Debye temperature of the acoustic phonon branch; $\gamma =$ high-temperature Grüneisen constant; $\delta^3 =$ volume per atom; M = average atomic mass; $\kappa_{calc} =$ calculated thermal conductivity from Eq. (2.13); $\kappa_{exp} =$ measured thermal conductivity.

Element/ Compound	θ_a (K)	γ	δ (Å)	M (amu)	κ_{calc} (Wcm^{-1}K^{-1})	κ_{exp} (Wcm^{-1}K^{-1})
C	1450	0.75	1.78	12.01	16.4	30
Si	395	1.06	2.71	28.08	1.71	1.66
Ge	235	1.06	2.82	72.59	0.97	0.65
BN	1200	0.7	1.81	12.41	11.05	7.6
BP	670	0.75	2.27	20.89	3.59	3.5
BAs	404	0.75	2.39	42.87	1.70	
AlP	381	0.75	2.73	28.98	1.10	
AlAs	270	0.66	2.83	50.95	0.89	0.98
AlSb	210	0.6	3.07	74.37	0.77	0.56
GaP	275	0.75	2.73	50.35	0.72	1.00
GaAs	220	0.75	2.83	72.32	0.55	0.45
GaSb	165	0.75	3.05	95.73	0.33	0.4
InP	220	0.6	2.94	72.90	0.83	0.93
InAs	165	0.57	3.03	94.87	0.51	0.3
InSb	135	0.56	3.24	118.29	0.38	0.2
ZnS	230	0.75	2.71	48.72	0.40	0.27
ZnSe	190	0.75	2.84	72.17	0.35	0.19
ZnTe	155	0.97	3.05	96.49	0.17	0.18
CdSe	130	0.6	3.06	95.68	0.23	
CdTe	120	0.52	3.23	120.00	0.296	0.075

structure compound, MgO, can be categorized as a high-thermal-conductivity compound, with $\kappa \approx 0.6$ Wcm^{-1}K^{-1} at room temperature.

We turn next to the zincblende and diamond structure compounds; see Table 2.2. One very striking feature of these compounds is that the Grüneisen parameters tend to be much lower than those of the rocksalt structure compounds: in the zincblende and diamond structures the phonons are more harmonic. In fact, for some members of this family, e.g., silicon, some of the mode Grüneisen parameters are negative [11]. Recent lattice dynamics calculations of the mode Grüneisen parameters for diamond, silicon, and boron nitride have been carried out [11, 12, 13]; see Fig. 2.3. Here we clearly see that the longitudinal modes tend to have Grüneisen parameters near unity, and the transverse modes have smaller, and even negative, γs. Of course the important parameter is the average value of the square of γ, and this is indicated in the figures. We see that the resulting average γs for these zincblende and diamond structure compounds as derived from lattice dynamics calculations are consistent with those presented in Table 2.2, which in most cases

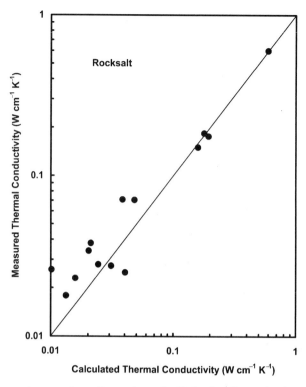

Fig. 2.2. Room-temperature thermal conductivity for the rocksalt compounds of Table 2.1 plotted against the thermal conductivity calculated from Eq. (2.13).

were derived from high-temperature thermal expansion data [1]. The necessity of a low γ for high-thermal conductivity is a recurring theme in this review. We see in Fig. 2.4 a very well-behaved relationship between measured and calculated room-temperature lattice thermal conductivities, spanning a range from $0.18\,\mathrm{Wcm}^{-1}\mathrm{K}^{-1}$ for ZnTe to $>30\,\mathrm{Wcm}^{-1}\mathrm{K}^{-1}$ for isotopically enriched diamond. Twelve members of this family of materials have or are expected to have thermal conductivity at room temperature in excess of $0.5\,\mathrm{Wcm}^{-1}\mathrm{K}^{-1}$ with several (diamond, BN, BP, Si, BAs, AlP, and GaP) exceeding $1\,\mathrm{Wcm}^{-1}\mathrm{K}^{-1}$. A more detailed description of the thermal conductivity of diamond is the subject of Chapter 7 in this book.

2.3.2 Wurtzite Crystal Structure

For the $n = 4$ wurtzite structure compounds CdS, ZnO, GaN, BeO, AlN, and SiC (Table 2.3 and Fig. 2.5) we again find excellent agreement between the calculated and measured room-temperature thermal conductivities, except

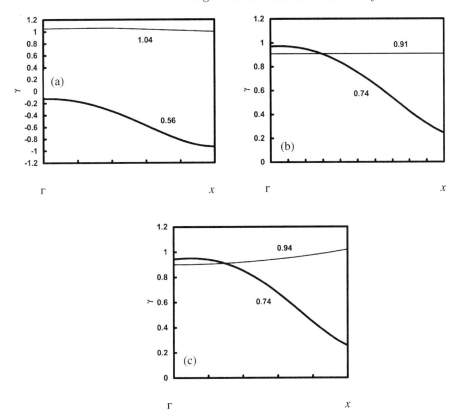

Fig. 2.3. Longitudinal (thin lines) and transverse (bold lines) mode Grüneisen parameters for (a) silicon, (b) diamond, and (c) boron nitride, with the average value $\langle \gamma_i^2 \rangle$ as indicated.

for the case of BeO, where the measured thermal conductivity exceeds the calculated value by about a factor of four. We note, however, that the value of $\gamma = 1.3$, which was derived from thermal expansion data [14], is significantly larger than that used for the other wurtzite compounds. Using a similar value of $\gamma = 0.75$ for BeO, in fact, improves greatly the agreement between the model and experiment. Further measurements or calculations of the Grüneisen parameter for BeO would be desirable. We note further that all of these compounds except CdS can be categorized as possessing high thermal conductivity according to our criterion. These crystals are undergoing significant development for their potentially useful electronic and optical properties; thus the last decade has seen a dramatic improvement in the availability and quality of single crystals of these wurtzites. Because of their technological potential, we will discuss the thermal conductivity of some of these crystals in more detail in Sect. 2.4.

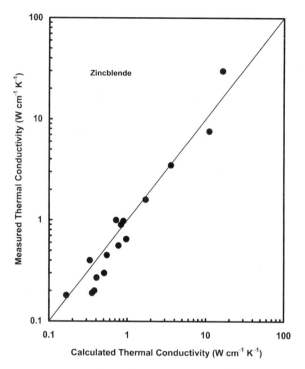

Fig. 2.4. Room-temperature thermal conductivity for the zincblende compounds of Table 2.2 plotted against the thermal conductivity calculated from Eq. (2.13).

Table 2.3. Calculated and experimental room-temperature thermal conductivity of several wurtzite ($n = 4$) compounds. θ_a = high-temperature Debye temperature of the acoustic phonon branch; γ = high-temperature Grüneisen constant; δ^3 = volume per atom; M = average atomic mass; κ_{calc} = calculated thermal conductivity from equation (2.13); κ_{exp} = measured thermal conductivity.

Compound	θ_a (K)	γ	δ (Å)	M (amu)	κ_{calc} (Wcm^{-1}K^{-1})	κ_{exp} (Wcm^{-1}K^{-1})
SiC	740	0.75	2.18	20.0	4.45	4.9
AlN	620	0.7	2.18	20.49	3.03	3.5
GaN	390	0.7	2.25	41.87	1.59	2.1
ZnO	303	0.75	2.29	40.69	0.65	0.6
BeO	809	1.38/0.75	1.90	12.51	0.90/3.17	3.7
CdS	135	0.75	2.92	72.23	0.13	0.16

2.3.3 Silicon Nitride and Related Structures

Up to now we have discussed structures containing only two or four atoms per primitive unit cell. We will now consider briefly a few compounds with

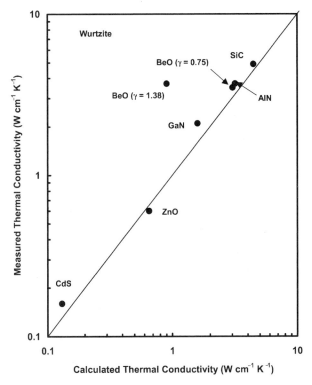

Fig. 2.5. Room-temperature thermal conductivity for the wurtzite compounds of Table 2.3 plotted against the thermal conductivity calculated from Eq. (2.13).

$n > 4$ that are potentially high-thermal-conductivity materials. One example is Si_3N_4. This compound assumes two crystal structures, known as the α and β phases, and is characterized by extreme hardness and toughness arising from predominantly covalent bonding [15]. Thus one might expect this compound to exhibit high thermal conductivity even though the crystal structure is not simple. Watari et al. [16] have reported the fabrication of hot-pressed, polycrystalline Si_3N_4 samples with thermal conductivity as high as $1.55\,\mathrm{Wcm^{-1}K^{-1}}$.

The α- and β-phases of Si_3N_4 are both hexagonal with $n = 28$ and $n = 14$, respectively [17]. Recently a third high-pressure phase, called γ-Si_3N_4, has been reported [18]. This phase crystallizes in the cubic spinel $MgAl_2O_4$ structure with $n = 14$. There are no thermal conductivity data on this structural modification in the literature. From recently published calculated phonon-dispersion curves [19] and thermal expansion data [20] one can estimate the average Debye temperature for these three structural modifications. Little information is available regarding the Grüneisen parameters of these compounds. He et al. [21] calculated $\gamma = 1.1$ for β-Si_3N_4. This value was derived from compressibility and bulk modulus data. Bruls et al. [22] report a high temperature $\gamma = 0.63$ for β-Si$_3$N$_4$, not too different from the value $\gamma = 0.72$ reported by Slack and Huseby [23]. In order to obtain an upper limit for the calculated thermal conductivity we will assume the smaller value of γ for both the α- and β-phases. We will use the value $\gamma = 1.2$ determined for the γ-phase [20], noting that this value is consistent with $\gamma = 1.4$ for the isostructural compound MgAl$_2$O$_4$ [24].

The necessary parameters for all three Si$_3$N$_4$ phases are collected in Table 2.4, along with the calculated thermal conductivities. We have also included results for MgAl$_2$O$_4$ that serve to verify the validity of Eq. (2.13) for these more complex crystal structures. We see that the calculated thermal conductivity for the α- and β-phases both exceed $1 \, \mathrm{Wcm^{-1}K^{-1}}$. The calculated value for β-Si$_3$N$_4$ suggests that even higher thermal conductivity than that measured by Watari et al. may be obtained in pure β-Si$_3$N$_4$ material. Reliable data on the Grüneisen parameters of these compounds would be very useful to verify the model for these compounds.

Ge$_3$N$_4$ also forms in the same α, β, and γ crystal structures [17, 25]. There are no experimental data on the thermal conductivity of these compounds; one would expect, however, in analogy to the Group IV semiconductors Si and Ge, that the heavier average mass of the germanium compounds will produce lower thermal conductivity than the silicon-based isostructures. From the available theoretical phonon-dispersion curves for the β-phase [26] and using the same value of γ as for β-Si$_3$N$_4$, we can make an estimate for the thermal conductivity of this compound; see Table 2.4.

Also very exciting from the point of view of high thermal conductivity are the predicted compounds C$_3$N$_4$ from the same α, β, and γ structural modifications [27, 28, 29] as well as a defective zincblende structure [30] and

Table 2.4. Calculated and experimental room-temperature thermal conductivity of several phases of Si$_3$N$_4$ and related compounds. n = number of atoms in the primitive unit cell; θ_a = Debye temperature of the acoustic phonon branch; γ = high-temperature Grüneisen constant; δ^3 = volume per atom; M = average atomic mass; κ_{calc} = calculated thermal conductivity from equation (2.13); κ_{exp} = measured thermal conductivity.

Compound	n	θ_a (K)	γ	δ (Å)	M (amu)	κ_{calc} (Wcm^{-1}K^{-1})	κ_{exp} (Wcm^{-1}K^{-1})
α-Si$_3$N$_4$	28	337	0.7	2.19	20.03	1.32	
β-Si$_3$N$_4$	14	485	0.7	2.18	20.03	2.61	1.55
γ-Si$_3$N$_4$	14	480	1.2	2.02	20.03	0.8	
γ-MgAl$_2$O$_4$	14	352	1.4	2.11	20.33	0.24	0.24
β-Ge$_3$N$_4$	14	243	0.63	2.31	39.11	0.65	
β-C$_3$N$_4$	14	~650	0.7	1.91	13.15	3.5	
Be$_2$SiO$_4$	42	316	1.02	2.06	15.73	0.35	
Zn$_2$SiO$_4$	42	236	0.52	2.318	31.83	1.29	
Zn$_2$GeO$_4$	42	186	0.31	2.367	28.19	2.17	

even a CN phase [31]. Several of these compounds, originally proposed by Cohen [32], are predicted to have bulk moduli rivaling that of diamond. Of course, many of the features favoring high hardness, such as short bond lengths and strong covalent bond character, give rise to high thermal conductivity. Thus it is likely, given the results on Si_3N_4 and the smaller mass of the carbon atom, that at least some of the C-N phases, if they exist, may possess thermal conductivities at least as high as their Si-based counterparts. Since the predictions of their existence, there have been numerous attempts [33, 34, 35, 36, 37] to synthesize various structural modifications of C_3N_4 and related phases, though it is debatable whether any has been demonstrated unequivocally [38]. We can make a rough estimate of the thermal conductivity of these compounds, although we do not have the luxury of lattice dynamical calculations of the phonon dispersion and phonon density of states. Rather, we make an estimate of the high-temperature Debye temperature from the theoretical bulk and shear moduli [39] using the method of Ravindran et al. [40]. The results are shown in Table 2.4. The predicted thermal conductivity of β-C_3N_4 exceeds that of β-Si_3N_4; if reasonably large crystals of the carbon nitrides become available it would be very interesting to study their thermal conductivity.

Be$_2$SiO$_4$ (phenacite) and Zn$_2$SiO$_4$ (willemite) also possess the β-Si_3N_4 structural modification but with two of the silicon atoms replaced by Be and Zn, respectively [41]. As with silicon nitride itself, these and other phenacites are typified by Grüneisen parameters on the order of or, in some cases, much less than, unity. Thus they could be potentially high-thermal-conductivity materials even though they have fairly large $n = 42$. From thermal expansion data the high-temperature limits of θ and γ have been determined [23] and θ_a calculated from Eq. (2.12). The estimated room-temperature thermal conductivity of these and related compounds are displayed in Table 2.5. Zn$_2$SiO$_4$ and Zn$_2$GeO$_4$ both display calculated thermal conductivity in excess of $1\,\mathrm{Wcm^{-1}K^{-1}}$; the case of Zn$_2GeO_4$, is very interesting because this compound has both a large n and large M. Again we see high thermal conductivity arising from a very small Grüneisen parameter; this suggests that looking for compounds with similarly low γ is another route for discovering high-thermal-conductivity materials. The tendency for a crystal to possess a low γ may be related to the openness of the structure [23]. This openness

Table 2.5. Calculated and experimental room-temperature thermal conductivity of some boron-containing compounds. Parameters are defined in Table 2.4.

Compound	n	θ_a (K)	γ	δ (Å)	M (amu)	κ_{calc} (Wcm^{-1}K^{-1})	κ_{exp} (Wcm^{-1}K^{-1})
B$_{12}$As$_2$	14	390	0.75	2.10	19.97	1.10	1.2
B$_{12}$P$_2$	14	481	0.75	2.06	13.69	1.38	0.38
B$_{12}$O$_2$	14	520	0.75	2.05	11.55	1.47	

allows more freedom of movement for the transverse phonon modes, and it is these modes that generally possess lower Grüneisen parameters [13]. Further detailed experimental and theoretical studies on the Grüneisen parameters and thermal conductivity of the phenacites and related structures would be very desirable to determine whether these compounds in fact possess small γ and large κ.

2.3.4 Icosahedral Boron Compounds

The element boron occurs in an α-structure, consisting of B_{12} icosahedra linked together with covalent bonds, and a β-structure consisting of B_{84} units [42]. Several boron-rich compounds also form as variations of the icosahedral B_{12} units [43]. From the point of view of high thermal conductivity, some of the most interesting of these are the compounds $B_{12}As_2$, $B_{12}P_2$, and $B_{12}O_2$. These compounds all have $n = 14$. The last of these, sometimes referred to as boron suboxide, was recently reported [44] to have a hardness exceeding that of cubic BN. Slack et al. [45] measured the thermal conductivity of a single crystal of $B_{12}As_2$ and an impure oligocrystalline $B_{12}P_2$ sample. Table 2.5 presents the necessary parameters to calculate the thermal conductivity. For $B_{12}As_2$ and $B_{12}P_2$, θ_a was calculated using Eq. (2.12) from θ-values estimated from the specific heat and elastic constants of similar boron compounds [45, 46, 47]; their γ-values were taken equal to that of β-boron [1]. We see that the model reproduces quite well the thermal conductivity of $B_{12}As_2$. As mentioned by Slack et al., the $B_{12}P_2$ they measured was neither a single crystal nor a very pure specimen, and examination of the temperature dependence of the thermal conductivity would suggest that its thermal conductivity at room temperature is partially limited by extrinsic scattering processes. It is likely that pure crystals of $B_{12}P_2$ and boron suboxide will have room-temperature thermal conductivity exceeding $1\,\mathrm{Wcm^{-1}K^{-1}}$.

In addition to these compounds, there are many other structures in the B-C-N triangle that exhibit hard or superhard behavior [48, 49], and it is possible that at least some of these may be high-thermal-conductivity materials. This is a rich field that currently is largely unexplored from the point of view of thermal transport and is deserving of further experimental and theoretical scrutiny.

2.3.5 Graphite and Related Materials

The form of carbon known as graphite is a hexagonal structure, $n = 4$, consisting of carbon atoms linked together in hexagons [50]. The C-C distance within the planes is $1.42\,\text{Å}$, nearly the same as that in benzene; the interplanar distance, on the other hand, is $3.40\,\text{Å}$. These differences reflect the very different nature of the bonding within a plane and between planes in graphite, with the former essentially a covalent sp^2-bonding arrangement and the latter a weak Van der Waals type of bonding. The thermal conductivity of graphite has been extremely well studied both experimentally and theoretically; for a more complete discussion the reader is referred to the review [51] and monograph [52] by Kelly, which include discussions of the influence of defects and various types of quasi-crystalline forms of this material. Here we only briefly consider graphite in its most perfect form, namely single crystals or highly oriented polycrystalline pyrolytic graphite.

Up to now, we have largely ignored the effects of anisotropy because for the crystals we have considered these effects are either absent or quite small. In graphite, however, the highly anisotropic nature of bonding manifests itself as an enormous anisotropy in the conduction of heat. Because of the crystal symmetry there are only two principle conductivities: that in the plane and that perpendicular to the plane, or along the so-called c-axis. Figure 2.6 shows composite curves of in-plane and c-axis thermal conductivity of graphite; these represent an average of many measurements that have been done over the last half century [53, 54, 55, 56, 57].

In the context of the simple model we have considered in this review, the anisotropy is due to the large difference in Debye temperature for phonon transport in the plane versus along the c-axis. These Debye temperatures can be estimated from a fit to the specific heat assuming a combination of "in-plane" and "out-of-plane" vibrations [58] and applying Eq. (2.12) to determine the Debye temperature of the acoustic modes.

Because of the strong intraplanar covalent bonding, we will assume for in-plane transport a Grüneisen parameter similar to that of diamond, while for out-of-plane transport we take $\gamma = 2$. The calculated thermal conductivity from Eq. (2.13) is shown in Table 2.6, and again we see that the simple model can account reasonably well for the magnitude of κ both perpendicular and parallel to the basal plane in graphite. A more complete theory of the thermal conductivity of graphite is based on the lattice dynamics models of in-plane and out-of-plane phonon modes (Komatsu [58]; Krumhansl and Brooks [61]) and the contribution of each of these to the basal plane and c-axis thermal conductivities. Extrinsic scattering mechanisms may also play an important role. The reader is referred to the book by Kelly [51] for further details.

Table 2.6. Thermal conductivity of graphite and BN in the basal plane (xy) and perpendicular to the c-axis (z).

Compound	n	θ_a (K)	γ	δ (Å)	M (amu)	κ_{calc} (Wcm^{-1}K^{-1})	κ_{exp} (Wcm^{-1}K^{-1})
Graphite-xy	4	1562	0.75	2.05	12.01	27	10–20
Graphite-z	4	818	2	2.05	12.01	0.5	0.06
BN-xy	4	1442	0.75	2.05	12.40	22	2–3
BN-z	4	755	2	2.05	12.4	0.4	$\sim$0.02

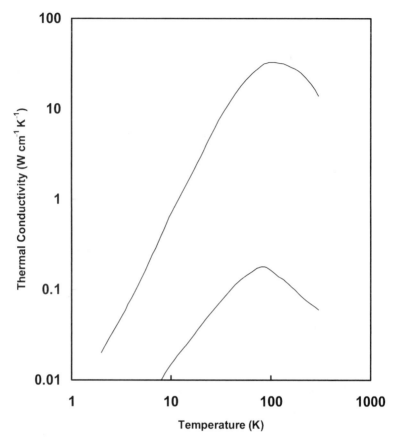

Fig. 2.6. Thermal conductivity of highly oriented graphite parallel (upper curve) and perpendicular (lower curve) to the basal plane.

Analogous to graphite, there exists a hexagonal form of boron nitride [62]. Measurements of the thermal conductivity have been made on sintered compacts with crystallite sizes on the order of 1000 Å or less [63]. We see (Table 2.7) again a reasonable agreement with the model using parameters derived in a similar fashion to those of graphite. Again, as in the case of graphite, a more complete theory of thermal conductivity in this hexagonal structural modification would take into account the details of the lattice dynamics of this structure.

Recently, many other forms of carbon, including fibers, sheets, C60, graphene sheets, and nanotubes, have been demonstrated or predicted. Some of these have or are expected to have high thermal conductivity. Chapter 8 of this volume is devoted to the thermal conductivity of carbon nanotubes and the reader is referred to it for further information.

2.4 Thermal Conductivity of Wide-Band-Gap Semiconductors: Silicon Carbide, Aluminum Nitride, and Gallium Nitride

We have seen that among the select group of materials with high thermal conductivity are the Group IV and Group III–V wide-band-gap semiconductors SiC, AlN, and GaN. Because of their wide gap, high-saturation electron velocities, and high thermal conductivity, these and related compounds have undergone significant development over the last decade for optoelectronic, high-frequency, high-temperature, and high-power device applications [64]. There has thus been an increase in availability of high-quality single crystals. Because of the importance of the thermal conductivity for many of these applications, we will look at these compounds in a little more detail in this section. Our emphasis is on the thermal conductivity of nearly defect-free single crystals and the influence of low levels of defects and impurities; Chapters 5 and 6 address the interesting and important subject of polycrystalline ceramics of SiC and AlN.

SiC was the earliest of this trio to undergo development as a substrate and active material for electronics applications. Much earlier, however, Slack [65] presented the first, and for many years the only, detailed characterization of the thermal conductivity of SiC single crystals and provided the first identification of this compound as a high-thermal-conductivity material. Slack noted that electrically active impurities had a noticeably stronger effect on the thermal conductivity than neutral impurities. Burgemeister et al. [66] studied several n- and p-type single crystals in the region around room temperature and showed that the thermal conductivity displayed a strong dependence on carrier concentration. Morelli et al. [67] studied several single crystals of different electron concentrations as a function of temperature. These results showed that samples with higher electron concentrations not only had lower thermal conductivity, but assumed a quadratic, as opposed to a T^3, temperature dependence at low temperature, an effect they ascribed to scattering of phonons by electrons in an impurity band. Müller et al. [68] measured the thermal conductivity of a single crystal from room temperature up to 2300 K. Some of these data are summarized in Fig. 2.7.

The data on SiC afford an example of how the relaxation-time approximation and the Debye model may be used to understand the magnitude and temperature dependence of the thermal conductivity. Theoretical fits of the lattice thermal conductivity may be performed using the standard expression (2.6). The phonon-scattering relaxation rate τ_C^{-1} can be written as:

$$\tau_C^{-1} = \frac{v}{L} + A\omega^4 + B\omega^2 T \exp\left(-\frac{\theta_D}{3T}\right) \qquad (2.14)$$

where the first term on the right-hand side represents scattering of the crystal boundaries with an effective crystal diameter L; the second term describes any

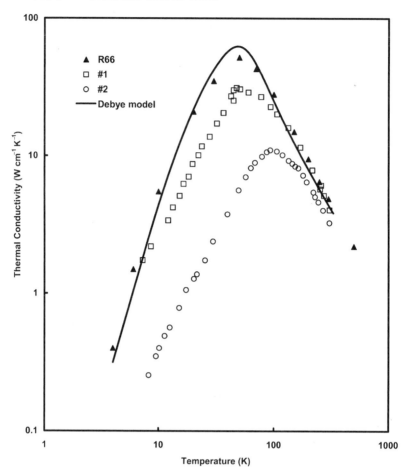

Fig. 2.7. Thermal conductivity of various single crystals of SiC. R66: pure crystal Slack [65]; #1 and #2: crystals [67] with electron concentrations of 3.5×10^{16} and 2.9×10^{18} cm^{-3}, respectively.

point-defect scattering that may be present in the crystal; and the third term represents intrinsic phonon-phonon Umklapp scattering. The data in Fig. 2.7 for the purest sample of SiC can be fit with this expression using a Debye temperature of $\theta_D = 800$ K.

The question of the influence of the electrical state of the sample on the thermal conductivity is important for the development of semi-insulating substrates for high-power electronic devices. Currently three-inch-diameter SiC substrates are commercially available and four-inch substrates have been demonstrated in the laboratory. Semi-insulating substrates are fabricated either by introducing vanadium during the growth process [69], thereby providing a deep level that traps free carriers, or by reducing as much as possible the presence of nitrogen during growth while providing "native" defects that can trap any remaining carriers [70]. In either case, to the extent that free carriers are eliminated, any reduction in the thermal conductivity below the "intrinsic" conductivity will be due to the presence of the trapping species. Studies of the thermal conductivity of silicon carbide containing these deep-level impurities would be very revealing in this regard.

Though not nearly as intense as the development of SiC substrates, GaN substrate development has accelerated dramatically over the last few years. As of this writing, two-inch wafers have been demonstrated and are becoming commercially available. Until recently, the only thermal-conductivity data available were those of Sichel and Pankove [71]. More recently, Pollak and coworkers have studied the local thermal conductivity of epitaxial layers of GaN using a scanning thermal microscopy technique. They found that the thermal conductivity of these layers depends strongly on the dislocation density, ranging from values as low as $1.3\,\mathrm{Wcm^{-1}K^{-1}}$ for high-dislocation density films to greater than $2\,\mathrm{Wcm^{-1}K^{-1}}$ for regions on films containing two orders of magnitude fewer dislocations [72]. Further studies by Pollak's group on n-type GaN layers showed [73] that the thermal conductivity also decreased strongly with increasing electronic concentration in the range 10^{17}–$10^{19}\,\mathrm{cm^{-3}}$. Slack et al. [74] recently reported the temperature-dependent thermal conductivity on a high-quality single crystal of GaN; these results are shown in Fig. 2.8 along with the earlier results of Sichel and Pankove. This single crystal had a room-temperature thermal conductivity of $2.1\,\mathrm{Wcm^{-1}K^{-1}}$ and the temperature dependence could be fit with Eq. (2.6) using a Debye temperature of approximately $525\,\mathrm{K}$. The large difference in the conductivities between this sample and that of Sichel and Pankove was attributed to the presence of oxygen in the latter sample.

The suggestion by Slack et al. that the difference in conductivities of these two GaN samples is due to the presence of oxygen was based on the well-documented studies of the thermal conductivity of the isostructural compound AlN. Although substrate development for this wide-band-gap semiconductor is still in its nascent stage [75], some information on the thermal conductivity is available in the literature. Slack et al. [76] studied several single crystals and found large differences in thermal conductivity that seemed to depend on oxygen content. A sample that was nearly free of impurities and defects had a room-temperature thermal conductivity of $3.5\,\mathrm{Wcm^{-1}K^{-1}}$, while those containing measurable quantities of oxygen impurity had a lower conductivity characterized by a depression or dip in the curve as a function of temperature. Some of these results, along with more recent results of Slack et al [74] on a sample containing about 1000 ppm oxygen, are shown in Fig. 2.9.

In order to gain a deeper understanding of the influence of oxygen on the thermal conductivity of AlN, it is useful to understand the kinetics of oxygen impurities in this compound. Oxygen in the aluminum nitride lattice has its origin in small amounts of Al_2O_3 dissolved in the AlN grains. At high

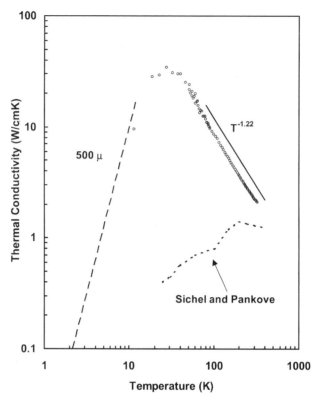

Fig. 2.8. Thermal conductivity of GaN as a function of temperature. Open points: data of Slack et al. [74]; lower dotted line: results of Sichel and Pankove [71]. Dashed line shows low-temperature boundary limit, assuming a crystal dimension of 500 microns.

temperatures, dissolution of Al_2O_3 occurs according to the reaction:

$$Al_2O_3 \longrightarrow 2Al + 3O_N + V_{Al}, \qquad (2.15)$$

where the subscripts N and Al, respectively, indicate that the O atoms occupy the nitrogen site and the vacancy occurs on the aluminum site. Thus, the presence of oxygen in the AlN lattice is always accompanied by the presence of vacancies, with the oxygen-vacancy ratio of 3:1. This is because the Al/O ratio in Al_2O_3 is 2:3 and the Al/N ratio in aluminum nitride is 1:1. The presence of an impurity (in this case oxygen) or a defect (in this case the vacancy) in an otherwise perfect aluminum nitride lattice will cause a reduction in thermal conductivity. This reduction has been well studied [2] and arises due to differences in the mass and size of the impurity or defect. These differences cause a scattering of the heat-carrying lattice vibrations; the scattering rate for this process is proportional to the square of the difference in mass between the host atom and the impurity. The mass difference between oxygen and nitrogen

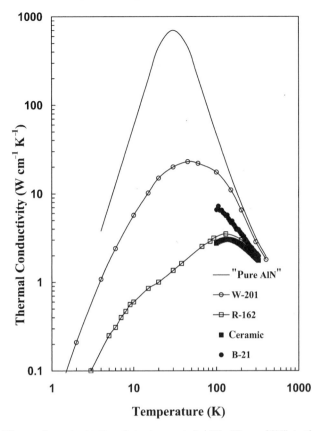

Fig. 2.9. Thermal conductivity of single-crystal AlN. "Pure AlN" is the calculated result for a crystal containing no impurities; samples W-201, R-162, and B-21 are single crystals with varying amounts of oxygen concentrations; see text. A ceramic sample is shown for comparison.

is not large; thus, the direct effect of oxygen on the thermal conductivity of the aluminum nitride is small. On the other hand, the fractional mass difference between a vacancy and aluminum is 100 percent, and this gives rise to a very large scattering rate. Thus, the lowering of the thermal conductivity by oxygen in AlN is really due to the presence of the vacancy on the aluminum site, which inexorably accompanies the less malevolent oxygen. The influence of oxygen is well described by the additive resistivity approximation [76]:

$$W_{\text{total}} = W_{\text{pure}} + \Delta W_1 \qquad (2.16)$$

were W_{total} is the measured thermal resistivity $(=1/\kappa_{\text{total}})$, W_{pure} is the thermal resistivity of a pure AlN crystal $(W_{\text{pure}} = 0.3\,\text{K cm W}^{-1})$, and ΔW is the increase in resistivity due to the presence of oxygen and is proportional to the oxygen concentration. For an oxygen concentration α by weight, the data of Fig. 2.3 yield $\Delta W = 110\alpha$ at room temperature. Thus, to obtain high thermal conductivity ($>0.5\,\mathrm{Wcm^{-1}K^{-1}}$), the oxygen concentration in AlN must be below about 1.5 percent. The influence of oxygen on the thermal conductivity of AlN is particularly important for the commercial manufacture of ceramic polycrystalline substrates of this material, as these ceramics are sintered using an oxide binder [77]. The thermal conductivity of AlN ceramics is discussed in detail in Chapter 5.

2.5 Isotope Effect in High Lattice Thermal Conductivity Materials

Since the early work of Pomeranchuk [78], it has been known that isotopes, due to their mass difference, can scatter phonons and decrease thermal conductivity. This effect was discussed also by Slack [79]. Geballe and Hull [80] provided unequivocal evidence for the influence of isotopes on the thermal conductivity with their experiments on natural abundance and isotopically purified germanium.

With the ready availability of isotopically purified source materials, the isotope effect has undergone reexamination over the last decade. Isotopically purified diamond [81, 82, 83, 84] displays a room-temperature isotope effect on the order of 40 percent. More recently, Asen-Palmer et al. [59] carried out a very thorough investigation of the isotope effect in germanium and showed that an isotopically purified sample had a 30 percent larger κ than natural abundance Ge. Very recently, Ruf et al. [85] reported an isotope effect in silicon of 60 percent at $300\,\mathrm{K}$, although the same authors [86] subsequently downgraded the magnitude to 10 percent.

The magnitude of the isotope effect in these materials is at first surprising because a simple estimate using the standard Debye theory of lattice thermal conductivity [3] (Eq. 2.6) yields increases in all cases of 5 percent or less. A more thorough and complete understanding of the isotope effect in these materials must recognize the importance of normal phonon-phonon scattering processes [87, 88, 89] within the context of the Callaway model. In the case of diamond it has been argued [87, 88, 89] that including the effect of normal phonon-scattering processes can explain the experimental result, although assuming infinitely rapid normal processes can only qualitatively fit the data [82].

Recently an extension of the Callaway model was provided by Asen-Palmer et al. [59], who successfully modeled the lattice thermal conductivity of Ge by not only using the Callaway formalism but also by considering the explicit mode dependence of the thermal conductivity and summing over one longitudinal (κ_L) and two degenerate transverse (κ_T) phonon branches:

$$\kappa = \kappa_L + 2\kappa_T, \tag{2.17}$$

where

$$\kappa_L = \kappa_{L1} + \kappa_{L2}. \tag{2.18}$$

The partial conductivities κ_{L1} and κ_{L2} are the usual Debye-Callaway terms given by:

$$\kappa_{L1} = \frac{1}{3}C_L T^3 \int_0^{\theta_L/T} \frac{\tau_C^L(x)x^4 e^x}{(e^x - 1)^2} dx, \tag{2.19}$$

$$\kappa_{L2} = \frac{1}{3}C_L T^3 \frac{\left[\int_0^{\theta_L/T} \frac{\tau_C^L(x)x^4 e^x}{\tau_N(x)(e^x - 1)^2} dx \right]^2}{\int_0^{\theta_L/T} \frac{\tau_C^L(x)x^4 e^x}{\tau_N^L(x)\tau_R^L(x)(e^x - 1)^2} dx}, \tag{2.20}$$

and similarly, for the transverse phonons,

$$\kappa_{T1} = \frac{1}{3}C_T T^3 \int_0^{\theta_T/T} \frac{\tau_C^T(x)x^4 e^x}{(e^x - 1)^2} dx, \tag{2.21}$$

$$\kappa_{T2} = \frac{1}{3}C_T T^3 \frac{\left[\int_0^{\theta_T/T} \frac{\tau_C^T(x)x^4 e^x}{\tau_N^T(x)(e^x - 1)^2} dx \right]^2}{\int_0^{\theta_T/T} \frac{\tau_C^T(x)x^4 e^x}{\tau_N^T(x)\tau_R^T(x)(e^x - 1)^2} dx}. \tag{2.22}$$

In these expressions, $(\tau_N)^{-1}$ is the scattering rate for normal phonon processes; $(\tau_R)^{-1}$ is the sum of all resistive scattering processes; and $(\tau_C)^{-1} = (\tau_N)^{-1} + (\tau_R)^{-1}$, with superscripts L and T denoting longitudinal and transverse phonons, respectively. The quantities θ_L and θ_T are Debye temperatures appropriate for the longitudinal and transverse phonon branches, respectively, and

$$C_{L(T)} = \frac{k_B^4}{2\pi^2 \hbar^3 v_{L(T)}} \tag{2.23}$$

and

$$x = \frac{\hbar\omega}{k_B T}. \tag{2.24}$$

Here ω is the phonon frequency and $v_{L(T)}$ are the longitudinal (transverse) acoustic phonon velocities, respectively.

The temperature dependence and the magnitude of the lattice thermal conductivity are determined by the temperature and frequency dependence of the scattering rates comprising $(\tau_N)^{-1}$ and $(\tau_R)^{-1}$, their coefficients, and the Debye temperatures and phonon velocities. The resistive scattering rate includes contributions from Umklapp processes, isotope scattering, and

Table 2.7. Percentage Increase in Room-Temperature Thermal Conductivity due to the Isotope Effect in Some Group IV and Group III–V semiconductors.

Element/ Compound	$\Delta\kappa/\kappa^*$ (%)	$\Delta\kappa/\kappa^{**}$ (%)	$\Delta\kappa/\kappa^{***}$ (%)
Ge	30	28	30
Si		12	60
C		23	35–45
SiC		36	
GaN		5	
BN		125	

* Model [59].
** Model [60].
*** Experimental results.

boundary scattering. Using the isotope scattering rate of Klemens [90] and appropriately adjusting the coefficients of the normal and Umklapp phonon-scattering rates, Asen-Palmer et al. [59] were able to quantitatively fit their experimental results over the entire temperature range of 10–300 K.

Recently, Morelli et al. [60] extended and modified this approach to model the isotope effect in diamond, silicon, germanium, silicon carbide, gallium nitride, and boron nitride. As mentioned previously, experimental data exist for the cases of diamond, silicon, and germanium. The model was able to account for the magnitude of the isotope effect all these semiconductors. Table 2.7 displays the measured and predicted isotope effect in these Group IV and Group III–V semiconductors at room temperature. Particularly note-worthy is the predicted magnitude of the isotope effect in boron nitride. This has its origin in the light atom masses and the natural abundance distribution of boron isotopes. Although boron nitride single crystals are extremely difficult to fabricate, it would be very desirable to study the isotope effect in this wide-band-gap semiconductor.

2.6 Summary

The intrinsic lattice thermal conductivity of crystalline solids near the Debye temperature can be understood on the basis of a simple model based on information that can be obtained from the crystal structure and lattice dynamics or phonon dispersion. The measured thermal conductivity of simple crystal structures such as rocksalt, zincblende, diamond, and wurtzite agree quite well with the predictions of the model. The model can be extended to other crystal structures to predict the thermal conductivity of new materials.

Several materials have been discussed that can be categorized as having high thermal conductivity. These include most of the simple zincblende, diamond, and wurtzite structure compounds. Various compounds possessing the hexagonal Si_3N_4 structure have been found or are predicted on the basis of this model to have high thermal conductivity, as have several compounds based on icosahedral boron structures. High lattice thermal conductivity compounds may be discovered in structures within the B-C-N or B-Si-N triangles and in similar triangles with oxygen substituted for nitrogen. This predicted high thermal conductivity would arise due to the strong covalent bonding and potentially low Grüneisen parameters of these structural modifications. The isotope effect may be used to increase the thermal conductivity of several wide-band-gap semiconductors, especially those containing boron.

Acknowledgments

The authors would like to acknowledge Dr. Joseph Heremans for useful discussions and a critical reading of the manuscript.

References

[1] G. A. Slack, *Solid St. Phys.* **34**, 1 (1979); G. A. Slack, *J. Phys. Chem. Solids* **34**, 321 (1973).

[2] P. G. Klemens, *Solid St. Phys.* **7**, 1 (1958).

[3] R. Berman, *Thermal Conduction in Solids* (Clarendon Press, Oxford, 1976).

[4] K. Watari and S. L. Shinde, Eds. *Mat. Res. Soc. Bull.* **26**, 440 (2001).

[5] S. Pettersson, *J. Phys. C: Solid St. Phys.* **20**, 1047 (1987).

[6] C. Domb and L. Salter, *Phil. Mag.* **43**, 1083 (1952).

[7] J. Callaway, *Phys. Rev.* **113**, 1046 (1959).

[8] G. Leibfried and E. Schlömann, *Nachr. Akad. Wiss. Göttingen II* **a(4)**, 71 (1954).

[9] C. L. Julian, *Phys. Rev.* **137**, A128 (1965).

[10] O. L. Anderson, *J. Phys. Chem. Solids* **12**, 41 (1959).

[11] C. H. Xu, C. Z. Wang, C. T. Chan, and K. M. Ho, *Phys. Rev. B* **43**, 5024 (1991).

[12] P. Pavone, K. Karch, O. Schütt, W. Windl, D. Strauch, P. Giannozzi, and S. Baroni, *Phys. Rev. B* **48**, 3156 (1993).

[13] G. Kern, G. Dresse, and J. Hafner, *Phys. Rev. B* **59**, 8551 (1999).

[14] Calculated from the data of G. A. Slack and S. F. Bartram, *J. Appl. Phys.* **46**, 89 (1975) and C. F. Cline, J. L. Dunegan, and G. W. Henderson, *J. Appl. Phys.* **38**, 1944 (1967).

[15] R. N. Katz, *Science* **84**, 208 (1980).

[16] K. Watari, K. Hirao, M. E. Brito, M. Toriyama, and S. Kanzaki, *J. Mater. Res.* **14**, 1538 (1999).

[17] R. W. G. Wyckoff, *Crystal Structures* (Interscience, New York, 1948), Vol. 2, p. 157.

[18] A. Zerr, G. Miehe, G. Serghio, M. Schwarz, E. Kroke, R. Riedel, H. Fue, P. Kroll, and R. Boehler, *Nature* **400**, 340 (1999).

[19] W.-Y. Ching, Y.-N. Xu, J. D. Gale, and M. Rühle, *J. Am. Cer. Soc.* **81**, 3189 (1998).

[20] J. Z. Jiang, H. Lindelov, L. Gerward, K. Stahl, J. M. Recio, P. Mori-Sanchez, S. Carlson, M. Mezouar, E. Dooryhee, A. Fitch, and D. J. Frost, *Phys. Rev. B* **65**, 161202 (2002).

[21] H. He, T. Sekine, T. Kobayashi, and H. Hirosaki, *Phys. Rev. B* **62**, 11412 (2000).

[22] R. J. Bruls, H. T. Hintzen, G. de With, R. Metselaar, and J. C. van Miltenburg, *J. Phys. Chem. Sol.* **62**, 783 (2001).

[23] G. A. Slack and I. C. Huseby, *J. Appl. Phys.* **53**, 6817 (1982).

[24] Z. P. Chang and G. R. Barsch, *J. Geophys. Res.* **78**, 2418 (1973).

[25] G. Serghiou, G. Miehe, O. Tschauner, A. Zerr, and R. Boehler, *J. Chem. Phys.* **111**, 4659 (1999).

[26] J. Dong, O. F. Sankey, S. K. Deb, G. Wolf, and P. F. McMillan, *Phys. Rev. B* **61**, 11979 (2000).

[27] D. Teter and R. J. Hemley, *Science* **271**, 53 (1996).

[28] A. Y. Liu and R. M. Wentzcovitch, *Phys. Rev. B* **50**, 10362 (1994).

[29] A. Y. Liu and M. L. Cohen, *Science* **245**, 841 (1989).

[30] J. Martin-Gil, F. J. Martin-Gil, M. Sarikaya, M. Qian, M. José-Yacamán, A. Rubio, *J. Appl. Phys.* **81**, 2555 (1997).

[31] M. Côté and M. L. Cohen, *Phys. Rev. B* **55**, 5684 (1997).

[32] M. L. Cohen, *Phys. Rev. B* **32**, 7988 (1985).

[33] C. Niu, Y. Z. Lu, and C. M. Lieber, *Science* **261**, 334 (1993).

[34] K. M. Liu, M. L. Cohen, E. E. Haller, W. L. Hansen, A. Y. Liu, and I. C. Wu, *Phys. Rev. B* **49**, 5034 (1994).

[35] H. W. Song, F. Z. Cui, X. M. He, W. Z. Li, and H. D. Li, *J. Phys. Cond. Matter* **6**, 6125 (1994).

[36] T.-Y. Yen and C.-P. Chou, *Solid St. Comm.* **95**, 281 (1995).

[37] Y. Chen, L. Guo, and E. Wang, *Phil. Mag. Lett.* **75**, 155 (1997).

[38] P. Ball, *Nature* **403**, 871 (2000).

[39] www.dirac.ms.virginia.edu/~emb3t/eos/html/final.html

[40] P. Ravindran, L. Fast, P. A. Korzhavyi, B. Johansson, J. Wills, and O. Eriksson, *J. Appl. Phys.* **84**, 4891 (1998).

[41] R. W. G. Wyckoff, *Crystal Structures* (Interscience, New York, 1948), Vol. 3, p. 133.

[42] R. W. G. Wyckoff, *Crystal Structures* (Interscience, New York, 1948), Vol. 1, p. 19.

[43] J. L. Hoard and R. E. Hughes, in *The Chemistry of Boron and Its Compounds*, ed. E. L. Muetterties (Wiley, New York, 1967), Chapter II.

[44] D. He, Y. Zhao, L. Daemen, J. Qian, T. D. Shen, and T. W. Zerda, *Appl. Phys. Lett.* **81**, 643 (2002).

[45] G. A. Slack, D. W. Oliver, and F. H. Horn, *Phys. Rev. B* **4**, 1714 (1971).

[46] G. A. Slack, *Phys. Rev.* **139**, A507 (1965).

[47] E. F. Steigmeier, *Appl. Phys. Lett.* **3**, 6 (1963).

[48] S. Veprek, *J. Vac. Sci. Technol.* **17**, 2401 (1999).

[49] P. Rogl and J. C. Schuster, eds., *Phase Diagrams of Ternary Boron Nitride and Silicon Nitride Systems* (ASM International, Metals Park, OH, 1992).

[50] R. W. G. Wyckoff, *Crystal Structures* (Interscience, New York, 1948), Vol. 1, p. 27.

[51] B. T. Kelly, in *Chemistry and Physics of Carbon*, ed. P. L. Walker, Jr. (Marcel Dekker, New York, 1969), Vol. 5, p. 119.

[52] B. T. Kelly, *Physics of Graphite* (Applied Science Publishers, London, 1981).

[53] G. A. Slack, *Phys. Rev.* **127**, 694 (1962).

[54] C. A. Klein and M. G. Holland, *Phys. Rev.* 136, A575 (1964).

[55] M. G. Holland, C. A. Klein, and W. B. Straub, *J. Phys. Chem. Solids* **27**, 903 (1966).

[56] A. de Combarieu, *J. Phys. (France)* **28**, 931 (1968).

[57] D. T. Morelli and C. Uher, *Phys. Rev. B* **31**, 6721 (1985).

[58] K. Komatsu, *J. Phys. Soc. Japan* **10**, 346 (1955).

[59] M. Asen-Palmer, K. Bartkowski, E. Gmelin, M. Cardona, A. P. Zhernov, A. V. Inyushkin, A. Taldenkov, V. I. Ozhogin, K. M. Itoh, and E. E. Haller, *Phys. Rev. B* **56**, 9431 (1997).

[60] D. T. Morelli, J. P. Heremans, and G. A. Slack, *Phys. Rev. B* **66**, 195304 (2002).

[61] J. A. Krumhansl and H. Brooks, *J. Chem. Phys.* **21**, 1663 (1953).

[62] R. W. G. Wyckoff, *Crystal Structures* (Interscience, New York, 1948), Vol. 1, p. 184.

[63] A. Simpson and A. D. Stuckes, *J. Phys. C* **4**, 1710 (1971).

[64] H. Morkoç, *Nitride Semiconductors and Devices* (Springer Verlag, New York, 1999).

[65] G. A. Slack, *J. Appl. Phys.* **35**, 3460 (1964).

[66] E. A. Burgemeister, W. von Muench, and E. Pettenpaul, *J. Appl. Phys.* **50**, 5790 (1979).

[67] D. T. Morelli, J. P. Heremans, C. P. Beetz, W. S. Yoo, and H. Matsunami, *Appl. Phys. Lett.* **63**, 3143 (1993).

[68] St. G. Müller, R. Eckstein, J. Fricke, D. Hofmann, R. Hofmann, R. Horn, H. Mehling, and O. Nilsson, *Materials Science Forum* **264–8**, 623 (1998).

[69] H. McD. Hobgood, R. C. Glass, G. Augustine, R. H. Hopkins, J. Jenny, M. Skowronski, W. C. Mitchel, and M. Roth, *Appl. Phys. Lett.* **66**, 1364 (1995).

[70] C. H. Carter, Jr., M. Brady, and V. F. Tsvetkov, *US Patent Number 6,218,680* (April 17, 2001).

[71] E. K. Sichel and J. I. Pankove, *J. Phys. Chem. Solids* **38**, 330 (1977).

[72] D. I. Florescu, V. M. Asnin, F. H. Pollak, A. M. Jones, J. C. Ramer, M. J. Schurman, and I. Ferguson, *Appl. Phys. Lett.* **77**, 1464 (2000); D. Kotchetkov, J. Zou, A. A. Balandin, D. I. Florescu, and F. H. Pollak, *Appl. Phys. Lett.* **79**, 4316 (2001).

[73] D. I. Florescu, V. M. Asnin, F. H. Pollak, R. J. Molnar, and C. E. C. Wood, *J. Appl. Phys.* **88**, 3295 (2000).

[74] G. A. Slack, L. J. Schowalter, D. T. Morelli, and J. A. Freitas, Jr., *Proc. 2002 Bulk Nitride Workshop*, Amazonas, Brazil (to appear in *Journal of Crystal Growth*).

[75] www.crystal-is.com

[76] G. A. Slack, R. A. Tanzilli, R. O. Pohl, and J. W. Vandersande, *J. Phys. Chem. Solids* **48**, 641 (1987).

[77] A. V. Virkar, T. B. Jackson, and R. A. Cutler, *J. Am. Ceram. Soc.* **72**, 2031 (1989).

[78] I. Pomeranchuk, *J. Phys. USSR* **4**, 259 (1941).

[79] G. A. Slack, *Phys. Rev.* **105**, 829 (1957).

[80] T. H. Geballe and G. W. Hull, *Phys. Rev.* **110**, 773 (1958).

[81] D. G. Onn, A. Witek, Y. Z. Qiu, T. R. Anthony, and W. F. Banholzer, *Phys. Rev. Lett.* **68**, 2806 (1992).

[82] T. R. Anthony, W. F. Banholzer, J. F. Fleischer, L. Wei, P. K. Kuo, R. L. Thomas, and R. W. Pryor, *Phys. Rev. B* **42**, 1104 (1990).

[83] J. R. Olson, R. O. Pohl, J. W. Vandersande, A. Zoltan, T. R. Anthony, and W. F. Banholzer, *Phys. Rev. B* **47**, 14850 (1993).

[84] L. Wei, P. K. Kuo, R. L. Thomas, T. R. Anthony, and W. F. Banholzer, *Phys. Rev. Lett.* **70**, 3764 (1993).

[85] T. Ruf, R. W. Henn, M. Asen-Palmer, E. Gmelin, M. Cardona, H.-J. Pohl, G. G. Devyatych, and P. G. Sennikov, *Solid St. Commun.* **115**, 243 (2000).

[86] T. Ruf, R. W. Henn, M. Asen-Palmer, E. Gmelin, M. Cardona, H.-J. Pohl, G. G. Devyatych, and P. G. Sennikov, *Solid St. Commun.* **127**, 257 (2003).

[87] K. C. Hass, M. A. Tamor, T. R. Anthony, and W. F. Banholzer, *Phys. Rev. B* **45**, 7171 (1992).

[88] R. Berman, *Phys. Rev. B* **45**, 5726 (1992).

[89] N. V. Novikov, A. P. Podoba, S. V. Shmegara, A. Witek, A. M. Zaitsev, A. B. Denisenko, W. R. Fahrner, and M. Werner, *Diamond and Related Materials* **8**, 1602 (1999).

[90] P. G. Klemens, *Proc. Roy. Soc.* **A68**, 1113 (1955).

3

Thermal Characterization of the High-Thermal-Conductivity Dielectrics

Yizhang Yang, Sadegh M. Sadeghipour, Wenjun Liu,
Mehdi Asheghi and Maxat Touzelbaev

It has been recognized that future improvements in performance and reliability of the microelectronic devices may only be possible through the use of new high-thermal-conductivity materials for thermal management in compact packaging systems. The diamond-like dielectric materials, in bulk form or thin film configurations are the likely choice, due to their high thermal conductivity and their excellent mechanical and electrical properties. However, the accurate thermal characterization of these materials has proven to be extremely challenging due to variations in fabrication processes and therefore their microstructures, as well as the practical difficulties in measuring small temperature gradients during the thermal characterization process. The variations in microstructure of these materials (e.g., CVD diamond) would manifest into *anisotropic, nonhomogeneous,* and *thickness-dependent* thermal properties that may vary by several orders of magnitude. As a result of these complications, a wide range of experimental techniques have been developed over the years, which may or may not be appropriate for thermal characterization of high-thermal-conductivity material of given microstructure and physical dimension. We will describe and critically review the existing thermal-characterization techniques for high-thermal-conductivity dielectric materials. In addition, we propose a number of techniques that are particularly tailored for accurate thermal characterization of diamond, silicon nitride (Si_3N_4), aluminum nitride (AlN), and silicon carbide (SiC) films and substrates. In each case, specific comments about the experimental technique and procedure, detailed description of the heat transfer process, and sensitivity analysis are provided.

3.1 Introduction

Continuous reduction in the size of electronic devices and systems and their process time scales requires fast removal of enormous heat for reliable performance. The lack of an efficient thermal-management strategy and system can often lead to overall system failures. This is not feasible except through use of high-thermal-conductivity materials as heat spreaders or heat sinks. A thorough thermal-conductivity evaluation of synthetic single crystals, combined with theoretical calculations, has revealed that most of the high-thermal-conductivity materials ($>100\,\mathrm{Wm^{-1}K^{-1}}$, at room temperature) are adamantine (diamond-like) compounds, for example, diamond, BN, SiC, BeO, BP, AlN, BeS, GaN, Si, AlP, and GaP [1]. The extensive research in recent years has concentrated on fabrication and characterization to obtain noble materials with improved mechanical or thermal properties. These efforts have resulted in development of new production or processing techniques, which have improved thermal conductivities of adamantine materials enormously (a few orders of magnitude, in some cases) and reduced their fabrication cost considerably. These developments have made the widespread use and commercialization of the passive diamond-like materials (in semiconductor devices and electronic systems) feasible and economically justified. For example, diamond is an excellent heat conductor and a good electrical insulator, rendering it ideal for passive applications. The high cost of synthesis at high temperatures and pressures, where diamond is thermodynamically stable, has been a barrier to the widespread commercial use of diamond in the past. However, evolution of the technology for low-pressure moderate-temperature chemical-vapor-deposition (CVD) of diamond on nondiamond substrate has removed this barrier.

Recent investigations [2]; [3] have led to the fact that, in addition to diamond itself, nitride and carbide, nonoxide diamond-like ceramics, demonstrate unique mechanical, electrical, and magnetic performances, compared to oxide ceramics. Accordingly, the significance of high-thermal-conductivity nonoxide ceramics has been recognized in many industrial fields. They have been increasingly used as heat spreaders for highly integrated circuits and optoelectronics, structural components for producing semiconductors, engine-related material components, etc. [4]. At present, SiC with measured thermal conductivity of 270–$360\,\mathrm{Wm^{-1}K^{-1}}$ and AlN are the commonly used ceramics for high-thermal-conductivity applications [1]. Due to the low reliability that arises from their poor mechanical properties, however, widespread use of SiC and AlN ceramics is still restricted. In addition, application of the SiC ceramics is further limited due to the low electric resistance and high dielectric constant. Silicon nitride (Si_3N_4) ceramic, on the other hand, is well known as a high-temperature structural ceramic with high strength and fracture toughness [5]. Compared with conventional processing techniques for single crystals and polycrystalline bulk ceramics, deposition process techniques have a strong potential for industrial application in terms of cost, scalability, and reliability. Chemical-vapor deposition (CVD) is a promising technique for producing bulk and film ceramics for a variety of applications, because it can yield materials with high purity and densification. Very recently, a significant increase in the thermal conductivity of polycrystalline films has been achieved. Several high-thermal-conductivity applications in the area of semiconductor processing, optics, electronics, and wear parts have been demonstrated [1]. Goodson [6] has categorized applications of the CVD diamond layers in electronic systems into three groups, called *generations* due to their varying stages of development. The first two generations consist of passive applications, and the third generation refers to the active application of diamond layers. The **first generation**, which has many industrial applications in production, uses thick (>100 μm) diamond plates for passive cooling of high-power electronic devices. Thermal resistance of the attachment material and distance between the active region and the diamond usually controls the effectiveness of using diamond as a heat spreader. The impact of diamond on the temperature rise in the device strongly depends on the time scale of the heating. The improvement for the case of a brief pulse heating can be larger than for the case of steady-state heating. However, using diamond may show no improvement if pulse duration is very short. The **second generation** of the CVD diamond layers, which is in applied research stage, is about deposition of the thin (<10 μm) diamond films within electronic microstructures to improve thermal conduction in the vicinity of the active semiconducting regions. The close proximity of the diamond to the active regions and direct deposition, needing no attachment material, will diminish the high temperature rise in the device. This will also promise a reduction in the maximum temperature rise in the device due to short pulse durations. Of course, deposition of the thin diamond films on nondiamond substrate results in a considerably lower thermal conductivity, especially at low temperatures, compared to the thick diamond plates. The **third generation** applications, currently under basic/applied research, use doped diamond as an active semiconductor in high-power electronic devices and thermistors.

It is well recognized that the definition of the thermal property and the thermal characterization of high-thermal-conductivity dielectric materials are extremely difficult mainly due to the fact that the variation in fabrication processes can strongly influence the material microstructures. In order to remove any ambiguity in this regard, a review of the microstructure of the CVD diamond, silicon nitride, aluminum nitride, and silicon carbide and its correlation with the observed *anisotropic, nonhomogeneous,* and *thickness-dependent* thermal properties of these materials are provided in Sect. 3.2. The available measurement techniques for thermal characterization of high-thermal-conductivity materials, with particular attention to the heating and thermometry techniques, are described in Sect 3.3. The role of time scale (Sect. 3.3.2) and the shape of heat source (Sect. 3.3.3) on the extracted thermal property and the extent of the heated (probed) zone in the transient techniques are demonstrated and discussed in detail. Section 3.4 discusses the steady-state techniques such as the DC heated bar method (Sect. 3.4.1), the film-on-substrate approach (Sect. 3.4.2), the suspended membrane technique (Sect. 3.4.3), and the comparator method (Sect. 3.4.4). A new structure is also proposed for fast and routine thermal characterization of high-thermal-conductivity dielectric films and substrates (Sect. 3.4.2). Sections 3.5 and 3.6 review the transient techniques. Frequency domain methods, such as the

Ångström thermal wave technique and the modified calorimetric method, are described in Sects 3.5.1 and 3.5.2, respectively. Applications of the 3ω technique for thermal characterization of (a) high-thermal-conductivity layers on low-thermal-conductivity substrates (Sect. 3.5.3); (b) anisotropic silicon nitride substrates (Sect. 3.5.4); and (c) spatially variable thermal conductivity of AlN substrates (Sect. 3.5.5) are also demonstrated. The time domain techniques, including laser heating (Sect. 3.6.1) and Joule heating (Sect. 3.6.2) and the thermal grating technique (Sect. 3.6.3) are reviewed and discussed in Sect. 3.6.

3.2 Microstructure of High-Thermal-Conductivity Dielectrics and Its Relevance to Thermal Transport Properties

Chemical-vapor deposition (CVD) is an attractive method for producing bulk and thin film materials for a variety of applications. The materials produced by CVD are theoretically dense, highly pure, and have other superior properties. By varying the process parameters, the CVD process can produce materials in a variety of forms such as single crystal, polycrystalline, or amorphous. Some examples of materials produced by CVD are diamond, SiC, Si, Si_3N_4, pyrolytic graphite and BN, ZnS, ZnSe, TiB_2, and B_4C [7].

3.2.1 CVD Diamond

Initial work on high-thermal-conductivity materials was carried out by Euken [8], who discovered that diamond was a reasonably good conductor of heat at room temperature. It has been determined [9] that intrinsic thermal conductivity of diamond is $2000\,Wm^{-1}K^{-1}$ at room temperature, much higher than that of either copper ($400\,Wm^{-1}K^{-1}$) or silver ($430\,Wm^{-1}K^{-1}$), which has the highest thermal conductivity of any metal at room temperature. The thermal properties of CVD diamond strongly depend on the microstructure of the material, which in turn is sensitive to the details of the process. The ratio of methane to hydrogen concentrations used during deposition strongly influences both the phase purity and the thermal conductivity of CVD diamond layers [10]. Diamond growth begins with the nucleation of the individual crystallites at random spots on the substrate, followed by competitive growth as the crystallites enlarge and merge. Those crystallites that happen to be oriented with their fastest crystallographic growth direction normal to the plane of the substrate eventually dominate, so that a strong columnar texture develops, with the long axis of the columnar grains being normal to the film; see Fig. 3.1. The average plane dimensions of the grains increase, more or less linearly, with distance z from the substrate. With such a microstructure, one would expect transport properties such as thermal conductivity to be anisotropic, nonhomogeneous, and thickness dependent [7], [12]. The local normal and lateral thermal conductivities are governed by phonon scattering on the grain boundaries [12]. The normal conductivity is usually greater than the lateral conductivity, due to the larger average separation distances between adjacent grain boundaries. Conductivities in both directions are very small in the regions close to the substrate due to very small grain sizes or very high grain boundary per unit volume. Graebner et al. [12], [13] measured the normal and lateral thermal conductivities of the CVD diamond films of different thicknesses, from which they extracted variations of the local thermal conductivities with the film thickness [14]. Their results show that both local normal and lateral thermal conductivities start from some small similar values close to the substrate and increase to the conductivity of

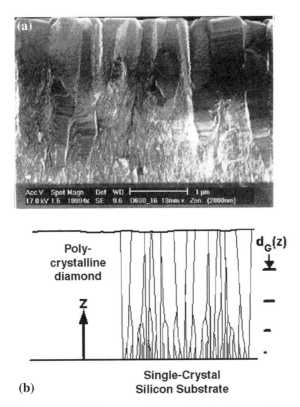

Fig. 3.1. (a) Cross-sectional electron micrograph of a diamond layer deposited at 800°C and nucleated using a bias voltage; (b) schematic of the grain structure in a micron-scale diamond layer deposited on silicon [11]. Reprinted with permission.

the gem-quality diamond with distance from the substrate ($>300\,\mu$m). Unlike the typical CVD-grown diamond films, the nanocrystalline films do not exhibit columnar growth and are often referred to in the literature as poor-quality diamond [15].

Figure 3.2 shows the range of the reported lateral and normal thermal conductivities of diamond layers with thickness less than $10\,\mu$m. The normal thermal resistances can be divided into two components, namely, the spatially average thermal resistance of the diamond layer, $R_f = d/k_f$, and the thermal boundary resistance at the diamond-silicon substrate interface, R_b. The interface thermal resistance is due to the formation of disordered microstructures, at the first stages of the crystal growth, which can dominate thermal conductivity in the diamond layers [11]. This additional thermal resistance can cause severe impediment to heat conduction for both packaging and device-cooling applications. As a result, measurements of the thermal boundary resistance at the interface of diamond-silicon substrate should be considered one of the most important components of the thermal characterization of diamond layers and substrates. The total thermal resistances of the diamond films deposited on silicon substrate, $R = R_f + R_b$, was estimated to be between 1.5×10^{-8} and $3.5 \times 10^{-8}\,\mathrm{m^2KW^{-1}}$ [16]. However, there is great scientific and technological interest in separating the contributions of the diamond layer and the thermal resistance at the interface. The measurement techniques for evaluation of the thermal boundary resistance of diamond layers are discussed in Sects 3.6.1 and 3.6.2.

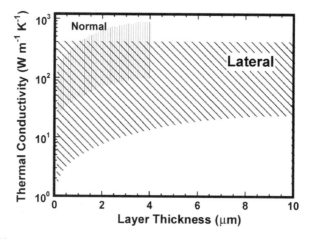

Fig. 3.2. The measured normal and lateral thermal conductivities of the diamond layers of thickness less than $10\,\mu$m; thermal conductivity is a strong function of the microstructure, and as a result, data are scattered [11].

3.2.2 CVD Silicon Nitride (Si$_3$N$_4$)

Silicon nitrides can be found in the forms of α- and β-Si$_3$N$_4$, where the former is unstable and converts to the more stable phase β-Si$_3$N$_4$. The β-Si$_3$N$_4$ formation has higher fracture toughness and therefore has been favored over α-Si$_3$N$_4$ ceramics. The thermal conductivity of β-Si$_3$N$_4$ ceramics has been increased over the past decade by a factor of three through development of densification and orientation technologies, application of sintering aids and raw powders, and firing at high temperatures (>2200 K). Details of the manufacturing process and the effect of microstructures on thermal conductivity of the silicon nitride have been thoroughly discussed and reviewed previously (e.g., [4]). Figures 3.3a–c show the microstructures of dense β-Si$_3$N$_4$ with Y$_2$O$_3$ additive fabricated by capsule-HIPing; gas-pressure sintering; and tape-casting and HIPing, respectively. Reported thermal conductivity values for these samples are $72\,\mathrm{Wm^{-1}K^{-1}}$ for capsule-HIPing [17]; $110\,\mathrm{Wm^{-1}K^{-1}}$ for gas-pressure sintering [18] and 155 and $65\,\mathrm{Wm^{-1}K^{-1}}$ along the tape casting and stacking directions, respectively, for the tape casting and HIPing. This brings us to the central point of our discussion that the orientation of β-Si$_3$N$_4$ grains results in the enhancement of the thermal conductivity and furthermore induces anisotropic thermal conduction in silicon nitride specimens [5]. It can be shown that the existing experimental techniques can be modified to obtain the thermal conductivity of anisotropic β-Si$_3$N$_4$ films; see Sect. 3.5.4.

3.2.3 Aluminum Nitride (AlN)

High thermal conductivity of the AlN ceramics combined with good electrical insulation has made it attractive for electronics packaging application. Thermal conductivity of sintered AlN has been enhanced over the years by resorting to effective sintering aid, development of high-purity fine powders and oxygen trapping into grains, and firing under a reduced-N$_2$ atmosphere. The

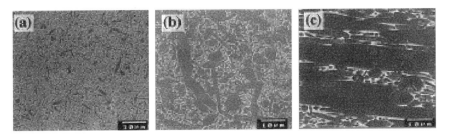

Fig. 3.3. Microstructure of the β-Si$_3$N$_4$ samples that are prepared using (a) capsule-HIPing at 2073 K, (b) gas-pressure sintering at 2273 K, and (c) tape casting with β-Si$_3$N$_4$ single-crystal particles as seeds and HIPing at 2773 K [4].

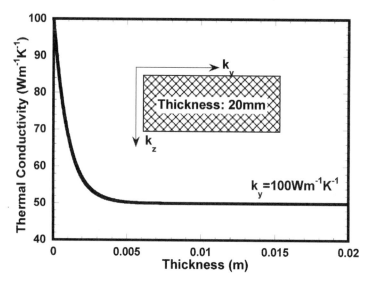

Fig. 3.4. Diffusion of the oxygen out of the AlN substrate induces an oxygen concentration gradient that can cause a spatial variation in thermal conductivity of the sample.

thermal conductivity of the pure single-crystal AlN is nearly $300\,\mathrm{Wm^{-1}K^{-1}}$ at room temperature, while significantly lower conductivities are reported for polycrystalline AlN due to random orientation of the grains, dissolved impurities (e.g., oxygen), and secondary phases with poor conductivities at the grain boundaries. The maximum solid solubility of oxygen in AlN at about $2000°C$ has been reported to be $2 \times 10^{21}\,\mathrm{cm^{-3}}$ or nearly 1.6 wt% [19]. It is suggested that the thermal conductivity of the single-crystal AlN ceramics would decrease linearly (from $300\,\mathrm{Wm^{-1}K^{-1}}$ at 0 wt% to $71\,\mathrm{Wm^{-1}K^{-1}}$ at 1 wt%) with added oxygen content [19]. On the other hand, impurities such as CaO or Y_2O_3 (1 wt%) are known to enhance densification of AlN ceramics, which in turn would improve the thermal conductivity by nearly a factor of two, up to a certain limit [20]. Extensive details about microstructure and composition of AlN ceramics and their thermal properties have been reported in the past [4], [20], [21] . Therefore, no further detailed information will be provided on this topic except the realization that there might be a strong oxygen concentration gradient in the sample during the manufacturing process. Buhr et al. [20] observed a strong gradient in oxygen content (0.5–$0.9\,\mathrm{wt\%}$) over the depth of $\sim$4 mm of a 9-mm AlN sample. Having in mind our previous discussions, this would induce a strong gradient in thermal properties of the AlN sample, as shown schematically in Fig. 3.4. In the past, a common practice was to cut the samples in pieces to obtain their local thermal-transport properties. We will, however, introduce a new approach here (Sect. 3.5.5) that enables us to obtain the thermal-conductivity gradients in AlN samples.

3.2.4 CVD Silicon Carbide (SiC)

SiC has polytypisms with numerous crystallographic modifications, some with better thermal conductivity than others. High-temperature treatments can lead to changes in the crystal structure, depending on the quality and type of raw powders and type of sintering aids. This effect may be used to enhance the polytype 6H, for example, which has high thermal conductivity relative to some other SiC polytypes. Densification is the method that has been used for SiC ceramics. The applied pressure during sintering and the effective additives are the controlling factors in obtaining the proper dense ceramics. The type of the sintering aids and the concentration of the aid elements dissolved in the grains remarkably affect thermal conductivity of the dense SiC. Four forms of SiC are commercially available: single-crystal, CVD, reaction-bonded, and hot-pressed. Their thermal conductivity values are 490, 300–74, 120–70 and 50–120 $Wm^{-1}K^{-1}$, respectively [7]. A small variation ($\sim$10%) in thermal conductivity of the relatively thick SiC substrates was observed across the sample thickness [7], such that no special consideration for thermal characterization of the bulk SiC is required.

3.3 Overview of the Measurement Techniques

Ever-increasing demands for packaging applications have encouraged the industry to invent new materials and improved deposition techniques for high-thermal-conductivity dielectric materials (e.g., CVD, sintering). These emerging process technologies yield materials that are polycrystalline and can vary considerably in properties and structure. Thin-film geometry, microcrystalline or amorphous structure of thin films, and the large number of potential defects due to the microfabrication process lead to inhomogeneity and anisotropic physical properties on a microscopic scale [22], [23]. As a result, thermal properties of materials in thin-film form in many cases differ strongly from those in bulk materials [24]. The term *thermal characterization* refers to diagnostic techniques that measure internal thermal resistance, thermal boundary resistance at the interfaces, lateral and normal thermal conductivities, and heat capacity of thin layers. A variety of thin-film thermal characterization techniques are available [24], [25], [26]; however, it is not always clear which technique is most appropriate for a given application. The time scale of the measurements and the geometry of the experimental structures influence the measured thermal-transport properties. If the film is nonhomogeneous, the region governing the signal can vary strongly depending on the measurement technique. For these reasons, it is possible to extract thermal property data for a given film that are substantially different from those governing the temperature distribution in a given device containing that particular film. It is therefore important that measurements be tailored to yield a specifically targeted property needed in the design process [27].

Recently, several outstanding review papers have been published on thermal-characterization techniques at micro- and nanoscales (e.g., [25], [26], [27], [28]) with more emphasis on low-thermal-conductivity materials. However, performing reliable thermal measurements on highly conducting material [29] is not a trivial task; it requires special consideration during the measurements. As a result, we only focus on thermal-characterization techniques that are more relevant to high-thermal-conductivity materials. We will pay particular attention to the techniques that are specifically tailored to measure the transport properties of diamond, AlN, SiC, and Si_3N_4 dielectrics. The experimental techniques for thermal characterization of high-thermal-conductivity dielectrics can be classified differently. It is common practice to categorize [30], [31] these techniques as follows: (a) *steady-state techniques*, which measure the heat flux necessary to maintain a fixed thermal gradient in the sample and are the only methods to obtain thermal conductivity directly; (b) *thermal wave (frequency domain) methods*, which allow measurement of the thermal diffusivity by determining the propagation constants of thermal waves in the sample and surrounding media; and (c) *pulsed (time domain) methods*, which deduce the diffusivity from the time required for a heat pulse to propagate through a section of the sample. Alternatively, one can classify the techniques under two more general steady-state and transient groups. With this classification, the thermal wave and pulsed methods will be under the transient division. The steady-state techniques result in direct evaluation of the thermal conductivities. The transient techniques (including the frequency and time-domain techniques), however, are based on the measurement of thermal diffusivity. Therefore, they require advance knowledge of the heat capacity for thermal conductivity evaluation. Fortunately, when adjusted for porosity, the heat capacity for a given material is not strongly sensitive to microstructure and, for most technologically relevant materials, the heat capacity per unit volume is known through measurements of bulk specimens [26]. Clearly, the sensitivities of these methods to the heterogeneity and anisotropy of high-thermal-conductivity dielectrics (e.g., CVD diamond) could be vastly different [32].

Heating and thermometry are the most essential actions in most thermal property measurements. Therefore, the measurement techniques in the preceding classifications are usually distinguished by their methods of heating and thermometry and by the temporal and spatial resolutions of the measurements. The measured properties are greatly affected by the measurement time scales, in particular the characteristic time scale of heating and the resolution of thermometry. We briefly review the existing heating and thermometry techniques and discuss the relevance of measurement time scales in precise determination of the type of properties measured by a given technique.

3.3.1 The Heating and Thermometry Techniques

The heat needed for measurements is usually induced by either Joule heating or absorption of thermal radiation. Each of these methods has advantages and disadvantages. For example, the accurate measurement of deposited heat in Joule heating is possible, but the absorbed heat in thermal radiation can't be quantified easily. On the other hand, fabrication of the special measurement structures on the sample surface and their electrical insulation, if needed, are required for Joule heating is more difficult than the surface preparation needed for the thermal radiation method, which is at most deposition of a metal film on the surface of the sample. The optical techniques are advantageous for thermal characterization of the novel materials whose chemical and structural stability during standard fabrication procedures are often not available. In this measurement technique, which is noncontact, the relative temperature response at the surface of the sample at different heating frequencies can be used to extract properties of the underlying layers.

There are different thermometry techniques, each of which has advantages and disadvantages. Electrical resistance thermometry is one of the most common and accurate temperature measurement techniques that can be precisely calibrated. Thermoreflectance thermometry uses the temperature dependence of the reflectivity to detect changes in the surface temperature of the sample. This technique has the advantage of being contact-free, needing minimum sample preparation, but the surface needs to be sufficiently reflective for the detector to collect the necessary radiation. Because of the small thermoreflectance coefficient of the metals at room temperature (10^{-4}–$10^{-6}\,\mathrm{K}^{-1}$) such measurements require averaging or lock-in detection to improve the signal-to-noise ratio [33].

3.3.2 Measurement Time Scale

Many of the techniques, such as 3ω and pump and probe, measure temperatures at heating locations, where the temperature rise is highest and easiest to measure. Figure 3.5 shows the surface temperature response as a function of frequency for a representative multilayer structure. In this structure, the heat is absorbed by the top metal layer (100 nm), which has a thermal resistance at its boundary with the next-highest thermal-conductivity diamond layer (5 μm). The femtosecond laser heating and thermometry technique is used to measure this thermal boundary resistance, as well as the thermal conductivity of the underlayer film. At longer time scales, heat reaches the thermal boundary resistance at the diamond-silicon substrate interface and then further diffuses into the substrate. Both nanosecond laser thermometry and Joule heating possess the relevant frequency range for these types of measurements. The electrical methods are more accurate but offer limited measurement frequency range. These low frequencies are suitable, though, for measuring the thermal boundary resistance at the diamond-silicon interface and the substrate thermal conductivity.

Measurements can also be done in the time domain, as in the thermal grating technique [34], [35] or in the laser flash technique [36]. Both of these methods measure the characteristic rise time at a location away from the point

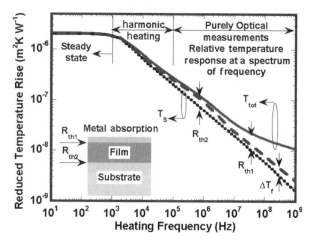

Fig. 3.5. Frequency range for the most common types of heating and thermometry. The layers, whose internal or boundary thermal properties govern the surface temperature response, change with the measurement frequency. Film thickness $d_f = 5\,\mu\mathrm{m}$ and substrate thickness $d_s = 300\,\mu\mathrm{m}$; thermal conductivity of film $k_f = 100\,\mathrm{Wm^{-1}K^{-1}}$, thermal conductivity of substrate $k_s = 150\,\mathrm{Wm^{-1}K^{-1}}$; boundary resistance at metal/film interface $R_{th1} = 1 \times 10^{-8}\,\mathrm{m^2KW^{-1}}$, boundary resistance at film/substrate interface $R_{th2} = 5 \times 10^{-8}\,\mathrm{m^2KW^{-1}}$; ΔT_f = temperature rise from the film; T_S = temperature rise from substrate; T_{tot}: total temperature rise of the structure.

of maximum heat flux. Due to the nature of the phase delay techniques, the measured property is the directional thermal diffusivity.

3.3.3 Impact of Geometry on Thermal Property Measurements in the Transient Techniques

Geometry of the heat source has a major impact on the type of properties, which can be extracted from the measurements; see Fig. 3.6 [27]. This effect can be demonstrated by examining the analytical solutions to the heat conduction equation. The geometry of induced heat flux in many experimental techniques can be approximated by a plane, line, or point source. Under a temporally harmonic, or periodic, heating source $Q_A(r,\omega)\exp[i(\omega t + \varphi_0)]$, where ω is the angular frequency and φ_0 is the initial phase, the solutions to the heat conduction in semi-infinite media are given as [37]:

$$\theta_{\mathrm{plane}}(r,\omega) = \frac{Q_A}{A}\frac{\exp(-pr)}{pk}, \tag{3.1a}$$

$$\theta_{\mathrm{line}}(r,\omega) = \frac{Q_A}{\pi L}\frac{K_0(pr)}{k}, \tag{3.1b}$$

$$\theta_{\mathrm{point}}(r,\omega) = \frac{Q_A}{2\pi}\frac{\exp(-pr)}{kr}, \tag{3.1c}$$

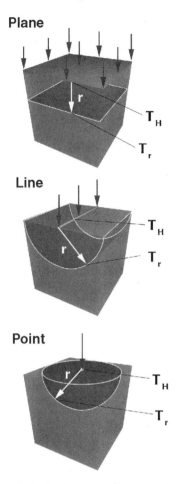

Fig. 3.6. The geometry of the heat source has a major impact on the type of properties that can be extracted from the measurements [27].

where A is the area of the plane heat source; r is the distance from the heat source; $p = (1 + i)/L_d$; $L_d = \sqrt{2\alpha/\omega}$ is the diffusion length; L is the length of the line source; k and α are the thermal conductivity and diffusivity; K_n is the modified Bessel function of order n; and $\theta(r, \omega) = T_A(r, \omega)\exp[-i\Delta\varphi(r, \omega)]$, where T_A and $\Delta\varphi$ are the amplitude and phase delay of the temperature, respectively. Knowing $\theta(r, \omega)$, temperature can then be found as:

$$T(r, \omega, t, \varphi_0) = \theta(r, \omega)\exp[i(\omega t + \varphi_0)]. \tag{3.2}$$

For convenience, the nondimensional parameters G and H may be introduced and defined as:

$$G = \frac{r}{\theta}\frac{\partial\theta}{\partial r} = \frac{\partial\ln(\theta)}{\partial\ln(r)} = \frac{\partial\ln(T_A)}{\partial\ln(r)} - i\frac{\partial\Delta\varphi}{\partial\ln(r)}, \tag{3.3a}$$

$$H = \frac{\partial \ln(\theta)}{\partial \ln(\omega)} = \frac{\partial \ln(T_A)}{\partial \ln(\omega)} - i\frac{\partial \Delta\varphi}{\partial \ln(\omega)}. \tag{3.3b}$$

The real and imaginary components of G and H can be determined experimentally by varying the position or frequency of the measurements. Analytical expressions for these parameters, only functions of the nondimensional variable pr, are given in Table 3.1. They can be fitted to experimental data to acquire a value for p and from that to obtain thermal diffusivity of the sample. There is no need for the absolute values of temperature and heat flux. The only requirements are sufficient bandwidth and the linear response of the thermometry technique over the temperature and frequency range encountered in the experiment. There is another class of techniques, however, which require absolute magnitudes of temperature and heat flux. These techniques use measurements at the heat source, $r = 0$. If $r = 0$, G and H become independent of p, which makes these techniques use measurements of temperature at distant locations, comparable to L_d, away from the heat source. Substituting the asymptotic values for $K_0(pr)$ and $\exp(-pr)$, when $pr \to 0$ in Eqs 3.1 will result in the equations for the functional dependence of the surface temperature on the measured angular frequency as:

$$\theta_{\text{plane}}(\omega) = \frac{Q_A}{A}\frac{1-i}{\sqrt{2}}k^{-1}\alpha^{\frac{1}{2}}[\omega^{-\frac{1}{2}}], \tag{3.4a}$$

$$\theta_{\text{line}}(\omega) = \frac{Q_A}{L}\frac{1}{2\pi}k^{-1}[-\ln(\omega)] + C, \tag{3.4b}$$

$$\theta_{\text{point}}(\omega) = Q_A\frac{1}{2\pi}\frac{(1+i)}{\sqrt{2}}k^{-1}\alpha^{-\frac{1}{2}}[-\omega^{\frac{1}{2}}] + C, \tag{3.4c}$$

where C is a constant independent of frequency. The results are summarized in Table 3.2. In the case of plane and point sources, either an in- or out-of-phase component may be used to find the corresponding combination of thermal properties.

Table 3.1. Mathematical expressions for analysis of experimental data directly governed by the sample's thermal diffusivity α. Thermal diffusivity is related to the parameter p. Experimental values of the functions G and H can be determined from temperatures measured at varying positions or frequencies. The assumed geometry is semi-infinite.

Source geometry	$G(r,\omega) = \dfrac{r}{\theta}\dfrac{\partial\theta}{\partial r}$	$H(r,\omega) = \dfrac{\omega}{\theta}\dfrac{\partial\theta}{\partial\omega}$
Plane	$-pr$	$-0.5(pr+1)$
Line	$-pr\dfrac{K_1(pr)}{K_0(pr)}$	$-0.5pr\dfrac{K_1(pr)}{K_0(pr)}$
Point	$-pr-1$	$-0.5pr$

Table 3.2. Mathematical expressions for the analysis of experimental temperature response obtained at the location of heat source ($r = 0$). The assumed geometry is semi-infinite [27].

Source Geometry	Plane	Line	Point
Extracted property	$k\alpha^{-0.5}$	k	$k\alpha^{0.5}$
Frequency dependence, $F(\omega)$	$\omega^{-0.5}$	$\ln(\omega)$	$\omega^{0.5}$
In-phase slope of temperature as a function of $F(\omega)$	$\dfrac{Q_A}{\sqrt{2}A}k^{-1}\alpha^{0.5}$	$-\dfrac{Q_A}{\left(\sqrt{2}\right)^2\pi L}k^{-1}$	$-\dfrac{Q_A}{\left(\sqrt{2}\right)^3\pi}k^{-1}\alpha^{-0.5}$
Out-of-phase slope of temperature as a function of $F(\omega)$	$-\dfrac{Q_A}{\sqrt{2}A}k^{-1}\alpha^{0.5}$	0	$-\dfrac{Q_A}{\left(\sqrt{2}\right)^3\pi}k^{-1}\alpha^{-0.5}$

3.4 Steady-State Techniques

Steady-state techniques induce a time-independent heat flux and measure the resulting temperature difference or distribution in the layer, from which thermal conductivity is then measured. Therefore, an adequate spatial resolution, which requires knowledge of at least two temperatures at precisely defined positions within the measurement structure, is particularly important for these techniques. Touzelbaev and Goodson [27] have outlined three requirements to be met by the geometry of the measurement in well-designed steady-state techniques. First, thermal resistance between the measurement positions should be controlled by or be strongly dependent on thermal conduction in the layer of interest. Second, this resistance should have very little or no dependence on thermal conduction to the environment once heat flows out of the measurement structure. And third, thermal resistance between the measurement locations has to be at least comparable to the total thermal resistance to the environment or heat sink in order to reduce measurement errors.

To measure the lateral thermal conductivity of the films, a long sample of uniform rectangular cross section (thickness d, width w) is usually used. The sample, which is thermally grounded at one end and fitted with a heater generating heat at a constant rate Q at the other end, can be either a free-standing layer (a suspended bridge) or a layer on a substrate with low thermal conductivity. The free-standing bridge consists of a layer without a substrate, but a film on a substrate is a layer-substrate composite. Assuming that all heat losses through the thermometer and heater lead wires are controlled and kept low by proper selection of materials, radiation and convection (or conduction) to the surrounding air and conduction to the substrate for the films on a substrate are the only paths of heat loss. If there is no heat loss to the air or substrate, all the heat generated by the heater Q will be conducted along the layer, causing a linear temperature distribution. In this case temperatures measured at two points, separated by a distance Δx, can be used to measure the lateral thermal conductivity as:

$$k_f = \frac{Q}{wd} \frac{\Delta x}{\Delta T}. \tag{3.5}$$

If surface losses to the surrounding area or substrate are not negligible, temperature distribution along the layer will be nonlinear and the equation governing its distribution will be:

$$\frac{d^2\theta}{dx^2} - m^2\theta = 0, \tag{3.6}$$

where $\theta = T - T_0$, T_0 is the base temperature, and $m = \sqrt{HP/k_f A}$ is a measure of the relative importance of heat conduction along the layer and heat losses (by radiation, convection, and conduction) from its surfaces. This formulation neglects any temperature change in the normal direction. The *healing length*, $L_H = 1/m$, is a characteristic length commonly used in the analysis that defines the length for which thermal resistance for the surface heat loss and conduction along the film are of similar orders of magnitude [38]. This defines a length of the film over which most of the temperature drop occurs. A large value of L_H is an indication of high thermal conductivity for the layer or low heat loss from its surfaces. $A(= w \times d)$, and $P(= 2w + 2d)$ are the cross-sectional area and the perimeter of the layer, respectively. $H(= H_{\mathrm{rad}} + H_{\mathrm{con}} + H_{\mathrm{sub}})$ is the total heat transfer coefficient for heat loss from the layer surfaces. It includes coefficients for radiation and conduction (or convection) exchange with the surroundings and heat loss by conduction to the substrate. For the films on substrate, heat loss to the substrate is dominant [38]. For the suspended bridges, however, this term does not exist, and the importance of heat loss to the surroundings depends mainly on the dimensions of the bridges.

For heat loss from the surfaces by radiation, we can write:

$$Q_{\mathrm{rad}} \doteq PL\varepsilon\sigma(T^4 - T_0^4), \tag{3.7}$$

where ε, the emissivity, takes a value between zero and one, depending on surface condition, and $\sigma(= 5.67 \times 10^{-8}\,\mathrm{Wm}^{-2}\mathrm{K}^{-4})$ is the Stephan Boltzmann constant. Using Eq. (3.7), an approximate equation for H_{rad} can be derived as:

$$H_{\mathrm{rad}} \approx 4\varepsilon\sigma T_0^3, \tag{3.8}$$

which is a good approximation if $[(T/T_0) - 1] \ll 2/3$. For measurements at room temperature $(T_0 = 300\,\mathrm{K})$ and with emissivity equal to one, $H_{\mathrm{rad}} = 6.1\,\mathrm{Wm}^{-2}\mathrm{K}^{-1}$. Obviously, this value will increase or decrease proportional to the temperature cube at other temperatures.

For most of the cases, a test is conducted in a vacuum so that there is no convection effect. For other cases, there will be a free convection heat transfer with a coefficient that usually has an order of magnitude of ten. However, because of the small dimensions of the layer, this coefficient will be much smaller, and in fact the heat transfer regime approaches that of molecular air conduction heat transfer.

To evaluate the importance of the heat transfer from the surface relative to that along the layer, we may compare their relevant thermal resistances:

$$R_{f,\text{cond}}/R_{f,\text{surf}} = \frac{Hp}{k_f A} L^2 = \left(\frac{L}{L_H}\right)^2, \tag{3.9}$$

where $R_{f,\text{cond}}$ and $R_{f,\text{surf}}$ are the thermal resistance for lateral heat conduction and combined radiation and convection or conduction from the surface, respectively. This ratio, and as a result heat loss from the surface relative to heat conduction along the layer, can be reduced by reducing the length L relative to the *healing length*, L_H. This can be achieved by either reducing L or increasing L_H. Considering the fact that the diamond-related materials generally have high thermal conductivities for testing long (especially thin) samples at room or higher temperatures, it is less likely to have a case with negligible surface heat loss. As a result temperature distribution will be nonlinear, which requires temperature measurements at more than two points.

3.4.1 The Heated Suspended Bar Technique

The DC heated-bar technique has been used to measure the lateral thermal conductivities of the layers of various materials. Graebner et al. [12] used this method to measure the lateral thermal conductivity of the CVD diamond films, of thickness 27.1 to 355 μm at room temperature, from which they extracted the local thermal conductivity. This technique was later used to extend the measurements to cover the wider temperature ranges of 0.15–7 K [39], 5–400 K [40], and 77–900 [41]. In this method, one end of the bridge is attached (clamped) to an isothermal heat sink at temperature T_0 in a vacuum chamber. The bridge is equipped with an electrical resistance heater (attached to or deposited directly on the surface) at its suspended end and a row of thermometers (thermocouples or resistive thermometers) along its surface to measure the local temperatures. The thermometers should be mostly within the L_H length, if $L_H < L$. Primary causes of experimental errors in this technique are conduction through the thermocouples and heater wires, which can be minimized by using very thin wires, and radiation from the heater [25]. Surface heat loss by molecular air conduction may be avoided by performing the experiment in a vacuum, although heat loss by surface radiation may still be important. Radiation heat transfer from the surface causes temperatures along the bridge to fall slightly from the linear distribution. Equation (3.6) governs this temperature distribution. Solving the equation with the boundary conditions $\theta = \theta_1 = T_1 - T_0$ at $x = 0$ and $\theta = \theta_2 = T_2 - T_0$, at $x = L$, temperature distribution will be:

$$\theta(x) = \frac{\theta_1 \sinh(m(L-x)) + \theta_2 \sinh(mx)}{\sinh(mL)}. \tag{3.10}$$

By changing θ_1, θ_2, and m one can find the best fit for this equation to the measured temperatures. The lateral thermal conductivity can then be calculated using the boundary condition on the heat flux at the heater, $k_f[dT/dx]_{x=L} = Q/A$ [31]. Selection of the temperature gradient at the heater will reduce the effect of radiation on the temperature gradient:

$$k_f = \frac{Q \sinh(mL)}{mA[\theta_2 \cosh(mL) - \theta_1]} \tag{3.11}$$

Clearly, the temperature gradient at $x = L$, and as a result k_f calculated from Equation (3.11), are very sensitive to the accuracy of the curve fit. The two-heater method with high sensitivity has been used to detect and correct for the presence of surface heat loss using only two instead of a row of thermometers [12], [40], [42]. The second heater, identical to the first one, is placed on the bridge near the thermal ground. In the absence of any heat loss from the surface, turning on each heater H_1 and H_2 individually, while the other is off, will result in a linear and uniform temperature distribution along the bridge, as shown by the solid lines in Fig. 3.7. The dashed lines represent the temperature distributions in the presence of radiation from the surfaces. The difference between the solid and dashed lines is an indication of the surface heat loss. To correct for the effects of surface radiation, one may add the temperature difference for the case when the heater H_2 is on to that of the case when the heater H_1 is on. In this way, the effect of surface radiation on the temperature distribution in the bridge is reasonably compensated for and, therefore (Eq. 3.5) for no surface radiation case, it can be used to measure the thermal conductivity. The distance between the heater and the nearest thermometer should be at least 5 times the thickness of the bridge to ensure a uniform distribution of the thermal current across the cross section of the sample in the vicinity of thermometers. At low temperatures, the mean free path of phonons, Λ, becomes comparable to the dimension of the layer, d. As a result, the heat transport around the heater element is ballistic and the heat diffusion equation is no longer applicable. However, at distances in the order of 2–3 Λ, or alternatively 2–3d, away from the heater, this effect is reduced considerably. By positioning the closest thermometer to the heater at a distance five times the film thickness, it will also be outside the ballistic effect region. The overall accuracy in the measurement of k_f is usually limited by the accuracy of the sample dimensions. For polished laser-cut diamond bridges, comparing with the as-grown CVD samples, better accuracies can be achieved [32].

The advantages of the DC heated-bar technique are [32] the ease of analysis due to one-dimensional heat conduction configuration and high accuracy

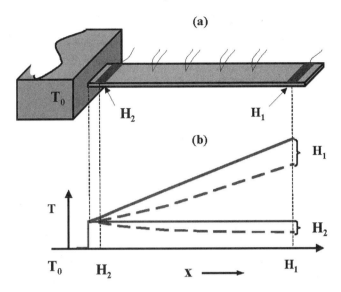

Fig. 3.7. (a) Experimental arrangement for the two-heater-bar technique for measuring the lateral thermal conductivity. Heater H_2 is used to test and correct for the surface heat loss. (b) For no surface heat loss, one expects a linear temperature distribution, as indicated schematically by the solid lines, for either H_1 or H_2 energized. Having significant surface heat loss, one expects the curved lines (dashed), as described in the text [40].

in the measurement of the steady temperature gradient. This technique is, however, very labor-intensive due to the small size of the samples, heaters, and thermometers. The heat losses from the heater through the connecting wires could become significant if the heater is not properly attached to the testing sample. In addition, it is susceptible to significant radiative or convective losses for poorly conducting materials, but it can be properly accounted for according to the procedure described previously [12], [40], [42]. Figure 3.8 can be used as a guideline to assess the error associated with the radiation loss for a given sample dimension and the best estimate of its thermal conductivity. For a given thermal conductivity value, each line represents the proper sample thickness and length for achieving less than 5% error due only to radiation loss. Measuring thermal conductivity of a film layer with a particular thickness on the left side of the corresponding line for choosing the sample length ensures negligible surface heat loss by radiation. However, for a sample length, which puts us on the right side of the lines, surface heat loss becomes important and should be considered. For example, given the dimensions and range of thermal conductivities in Asheghi et al.'s [43] experiments, the radiation losses would have been negligible even for layer thicknesses down to $1\,\mu$m.

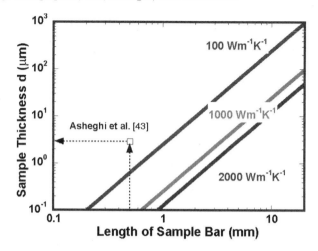

Fig. 3.8. A guideline to assess the error associated with the radiation loss as a function of the sample dimensions and its estimated (or measured) thermal conductivity.

3.4.2 The Film-on-Substrate Technique

Another DC heated method that has been used to measure the film's lateral thermal conductivity uses film-on-substrate configuration. The whole structure consists of the film layer (of thickness d and conductivity k_f), the underlying low-thermal-conductivity layer (of thickness d_0 and conductivity k_0), and a high-thermal-conductivity substrate. Due to very low thermal resistance in the path of heat flow to the substrate by conduction relative to that to the surroundings by radiation, most of the surface heat loss will be absorbed by the substrate, causing a very short healing length. While the heat losses by surface radiation and convection cannot be assessed accurately or controlled properly, the heat loss to the substrate through the low-thermal-conductivity material can be tailored and modeled more accurately. Asheghi et al. [38] used this method to measure the thermal conductivity of single-crystal silicon layers within the temperature range of 20–300 K. This idea can be used to design a test bed for measuring the lateral thermal conductivity of the diamond-like thin films by sandwiching them between two composite layers or shells (Fig. 3.9). Each layer consists of a thin (plexi) glass layer ($d_0 = 1$ mm and $k_o \approx 0.5\,\mathrm{Wm^{-1}K^{-1}}$) as the low-thermal-conductivity material and a thick (1 cm) copper layer ($k \approx 400\,\mathrm{Wm^{-1}K^{-1}}$), as the substrate (heat sink). The electric resistance heater and the thermometers are deposited on one of the glass layers and are covered with polyimids; therefore this setup can be used repeatedly for thermal conductivity measurement of different layers (Fig. 3.10). Some thermal grease, oil, or even water must be applied to improve the thermal contact between the sample and the glass. The whole structure should then be put under pressure to create a good contact between the sample surfaces and the glass, the heater, and the thermometers.

The thermal resistance associated with the adhesive layer will not play a role in the measurements due to the large thermal resistance of the glass layer and the high thermal conductivity of the sample under test. The second shell in fact serves two different purposes: preventing any possible heat loss from the top surface by gas conduction or radiation and making it possible to apply pressure for better contact.

Heat conduction in the multilayer structure shown in Fig. 3.10a can be approximated by the one-dimensional fin model (Fig. 3.10b) if the following three conditions are satisfied [38]:

1. The temperature variation across the substrate is small compared to the temperature drop across the low-thermal-conductivity layer.
2. The temperature variation in the film normal to the substrate (y direction) at any position x is negligible compared to that in the low-thermal-conductivity underlying layer. This condition is met if the ratio of normal thermal resistance of the film to that of the underlying layer is small, $(d/k_f)/(d_0/k_0) \ll 1$.
3. The lateral conduction in the underlying layer is negligible compared to that within the film. This condition is satisfied if the ratio of their lateral thermal resistances is small, $(d_0 k_0/d k_f) \ll 1$.

This problem now resembles the previous problem of one-dimensional heat flow in a suspended bridge. Simple fin equations and appropriate boundary conditions can very accurately predict solutions to the heat-diffusion equation in the multilayer system consisting of the high-thermal-conductivity layer, glass layer, and high-thermal-conductivity substrate. Therefore, $P = 2L_1$, $A = dL_1$, and $H = k_0/d_0$, and the equivalent definitions for m and healing length will then be $m = (2k_0/k_f dd_0)^{1/2}$ and $L_H = (k_f dd_0/2k_0)^{1/2}$.

Two fins are considered, one with $(-W < x < 0)$ and the other without $(L_2 > x > 0)$ heat-generation term (heater), as depicted in Fig. 3.10b. Boundary conditions include adiabatic ends at $x = -W$ and L_2 and continuity of temperatures and heat fluxes at $x = 0$. Temperature distribution for the fin without heat source, $L_2 > x > 0$, is given by

$$\theta(x) = T(x) - T_0 = C \cosh m(x - L_2), \qquad (3.12a)$$

where

$$C = \frac{Q \sinh(mW)}{WHP \sinh m(W + L_2)}. \qquad (3.12b)$$

Figure 3.11a shows the ranges of thermal conductivity and layer thickness that satisfy these conditions such that the lines A: $d = 0.01 \ (d_0/k_0) \ k_f$ and B: $d = 100 \ (d_0 k_0)/k_f$ would determine the regions where the one- or two-dimensional heat-conduction models are applicable.

Figure 3.11b shows the difference between the predicted values of $\Delta T = T_1(x_1 = 5d) - T_2(x_2 = 3L_H)$ as a function of k_f, thermal conductivity, from solutions of the 2-D heat-diffusion equation in the multilayer system and 1-D

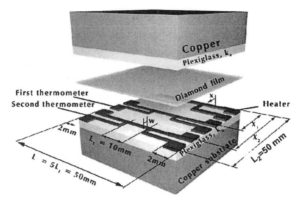

Fig. 3.9. Schematic of a test setup in which the sample is sandwiched between two identical substrates to avoid surface radiation.

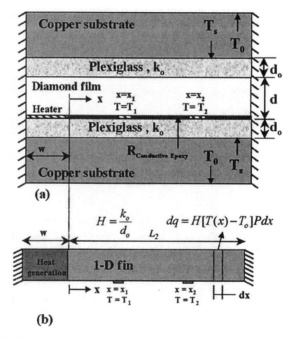

Fig. 3.10. (a) Cross section of the actual experimental setup, and (b) the simplified geometry used for the one-dimensional heat-conduction model in the high-thermal-conductivity layer.

fin model given by Eq. (3.12a). One should notice that the absolute values of the temperature $T_1(x_1)$ and $T_2(x_2)$ become irrelevant in the evaluation of the thermal-conductivity value. For $x_2 < 3L_H$, the differences between these two (1-D and 2-D) models are negligible. Results in Fig. 3.11b are given for two values of the thickness of the film layer and the low-thermal-conductivity

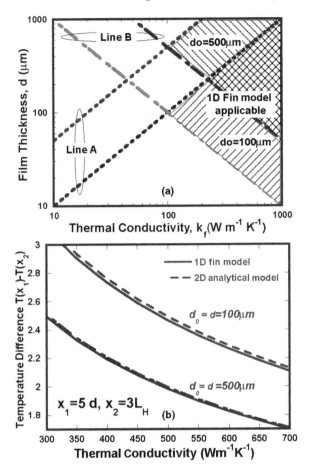

Fig. 3.11. (a) The range of the applicability of the one-dimensional model (Eq. 3.12a) is shown for $d_o = 100,500\,\mu\text{m}$ and $k_o = 1\,\text{Wm}^{-1}\text{K}^{-1}$. (b) Sensitivity curves for the temperature difference, as a function of the thermal conductivity and thickness of the sample, d, and the thickness of the low-thermal-conductivity film, d_o.

layer. Given the accuracy in the measurement of the temperature difference at locations x_1 and x_2, one can obtain the uncertainty in thermal conductivity of the layer under investigation.

3.4.3 The DC Heated Suspended Membrane

Another version of the DC heated-bar technique is the suspended membrane, which is designed to reduce the effects of surface radiation to a negligible level and to make fabrication of the suspended thin films feasible.

As is seen from Eq. (3.9), a way to minimize radiation effects is to decrease the length L. To achieve this goal, Graebner et al. [44] made very thin samples of diamond, deposited on silicon, free standing by etching a rectangular window entirely through the silicon. The remaining substrate then served as a natural support and heat sink for the film. The choice of relative dimensions of the window is crucial for reliable measured temperatures that have not been distorted by the surface radiation. The window has length L and width L_2 with the heater deposited onto the surface of the diamond film along its midline. The thermal resistances in the paths of heat conduction in film from the heater to the silicon substrate and radiation heat loss from the surface to the surroundings can be approximated as $R_{f,\text{cond}} \approx L_2/(4k_f L_1 d)$ and $R_{f,\text{surf}} \approx 1/(2L_1 L_2 H)$, respectively. The ratio of these two terms then becomes:

$$\frac{R_{f,\text{cond}}}{R_{f,\text{surf}}} \approx \frac{HP}{k_f A}(L_2/2)^2 = \left(\frac{L_2/2}{L_H}\right)^2, \qquad (3.13)$$

where $A = d \times L_1$, $P = 2L_1$, d is the film thickness, and H is the total heat transfer coefficient from the surface. As is seen, $L_2/2$, comparable to the length of the suspended bridge, plays a major role in determining the importance of the surface radiation. L_2, the width of the window, is usually kept small compared to the length of a free-standing bridge. For measuring the thermal conductivity of a film with $d = 1\,\mu\text{m}$, $k_f = 100\,\text{Wm}^{-1}\text{K}^{-1}$ at room temperature with $H = 6.1\,\text{Wm}^{-2}\text{K}^{-1}$, we find $L_H = 2.9\,\text{mm}$. Therefore, if we choose a window with width $L_2 = 2\,\text{mm}$, we have committed about 12% error for neglecting the surface radiation.

Circular membranes are also used to measure thermal conductivity of the large-area samples with irregular outlines. Jansen et al. [45] and Jansen and Obermeier [46], [47] performed finite element simulations to find temperature distributions in the heated rectangular and circular membrane and bar-structure micromechanical devices in order to find the best heater and thermoresistor pattern and to determine the effects of thermal radiation for the lateral thermal-conductivity measurements. The thermal conductivities of the CVD diamond films of different microstructures and thicknesses were measured over a temperature range of -195 to $300°\text{C}$, using the Joule-heating and electrical-resistance thermometry in these structures. The unusual shape of the suspended membrane, however, makes the data extraction procedure rather difficult. Graebner et al. [44] used a somewhat similar approach to measure the lateral thermal conductivity ($k = 190\text{–}600\,\text{Wm}^{-1}\text{K}^{-1}$) of suspended diamond membranes of thickness between 2.8 and 13.1 μm. The heater at the center of the suspended layer establishes a temperature distribution, which is a strong function of the membrane aspect ratio. For the given membrane dimensions ($2 \times 4\,\text{mm}^2$), the temperature distribution was found to be two-dimensional. Several thermocouples at the center of the structure and along the direction of the temperature gradient were used for thermometry at different locations. This method has the advantages that (1) only a relatively small sample is needed, (2) effects of thermal radiation are avoided, and (3) only modest expertise in thin film deposition and patterning is required. The thermal-conductivity data were estimated using a 2-D heat-conduction equation in the suspended membrane with the $T = T_0$ boundary condition at the base. The uncertainties in the positions of the thermocouples (typically $15\,\mu\text{m}$) were considered to be one source of the errors in this measurement. The temperature at the base of the membrane was higher than that of the substrate due to the large thermal resistance at the interface of the diamond-silicon substrate. This would introduce some error in this measurement and make the measurements extremely tedious.

Asheghi et al. [43] designed a structure that significantly simplifies the measurement procedure by increasing the aspect ratio of the membrane from $2:1$ [44] to $13:1$. Figure 3.12a shows a schematic of the experimental structure used to measure the lateral thermal conductivity of the doped silicon layers. The lateral dimensions of the suspended membrane are $1 \times 10\,\text{mm}^2$. The aluminum heater and thermometers are extended over the entire length of the membrane, but the power generated in the heater and the temperatures at points A and B are measured at the center of the suspended membrane within a region with lateral dimensions of $1 \times 1\,\text{mm}^2$. This is achieved by measuring the voltage drops in the aluminum bridges over the extent of

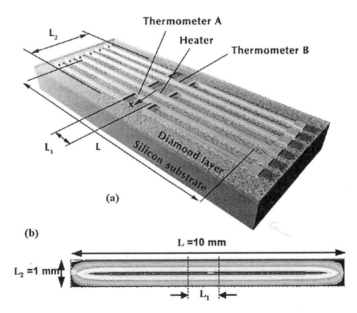

Fig. 3.12. (a) The experimental structure used to measure the lateral thermal conductivity of the SOI silicon device layer; (b) temperature distribution in the suspended membrane, which shows that the temperature distribution in the x-direction is one-dimensional.

the measurement section, L_1. This ensures one-dimensional heat conduction along the layer in the x-direction as verified by finite element calculations (Fig. 3.12b). The heater is located at the center of the suspended membrane. During the measurement, heat is generated by electrical current sustained in the aluminum line resulting in a linear temperature distribution along the $\langle 110 \rangle$ crystallographic direction. The temperatures at two locations above the silicon layer are detected using electrical-resistance thermometry in the patterned aluminum bridges A and B. The heat is conducted predominantly along the suspended membrane in the x-direction. Conduction to the surrounding air, conduction along the aluminum bridges, and radiation to the environment are negligible. The silicon layer thermal resistance is at least two orders of magnitude less than the thermal resistance due to surface radiation and three orders of magnitude less than the thermal resistance due to conduction along the heater and thermometer legs. The measurements are performed in a vacuum in order to minimize the heat conduction to the surrounding air. The thermal conductivity of the silicon layer is extracted using

$$k_f = \frac{(Q/2)}{A(\Delta T/\Delta X)}, \tag{3.14}$$

where $Q = I \times \Delta V$ is the power dissipation in the heater and I and ΔV are the current and voltage difference across the length L_1, respectively. Separation distance between bridges A and B is $\Delta x = 390\,\mu\text{m}$ and their temperature difference is ΔT. The cross-sectional area for heat conduction is $A = d \times L_1$. Figure 3.13 shows the thermal-conductivity data for the 3-μm-thick free-standing silicon layers, doped with boron and phosphorus at concentrations

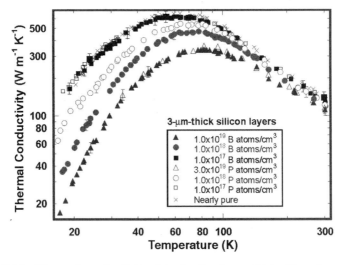

Fig. 3.13. Thermal conductivity data for the 3-μm-thick silicon layers [43].

ranging from 1×10^{17} to $3 \times 10^{19} \, \mathrm{cm^{-3}}$, measured at temperatures between 15 and 300 K [43].

3.4.4 The Comparator Method

The thermal comparator method was developed [48] and used extensively to measure thermal conductivity of bulk materials. The classical comparator method has certain restrictions, which limit its application to bulk materials with reasonably high hardness and low thermal conductivity. This method was then modified to measure the normal thermal conductivity of the low-thermal-conductivity dielectric films on substrates [49]. In this method, a heated probe is used and the temperature difference between the body of the probe and the tip that makes contact with the sample is measured (Fig. 3.14). During measurement, while keeping the temperatures of the copper block and the sample constant, the finger is pressed against the sample with a controlled force and the steady-state thermocouple voltage is recorded. Using the calibration curve, this voltage is then converted to apparent conductivity, which is a combined measure of the thermal resistances of the film, boundaries (at fingertip-sample, and sample-substrate interfaces), and substrate. The calibration curve is generated using bulk samples of the known thermal-conductivity materials.

There are difficulties in using the classical comparator method, particularly for thermal characterization of the high-thermal-conductivity films. Some of these problems and the solutions adopted in further modification of the method to make it usable for measuring the thermal conductivity of the CVD diamond films [50] are discussed here. The sample temperature approaches the probe temperature, using the classical comparator method to characterize the high-thermal-conductivity films. In this situation, heat transfer from the probe seems to be controlled by convection from the sample surface instead of sample thermal conductivity. To prevent this from happening, heat is removed from the base by active cooling to keep its temperature lower than the probe tip and the ambient temperatures.

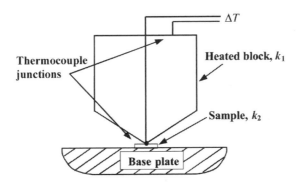

Fig. 3.14. Schematic of the comparator measurement technique.

The area of contact between the probe tip and the sample surface is meant to be the primary path for the flow of heat from the heated probe to the sample and therefore should be kept constant during calibration and measurement. Thus the proper compression force for each sample material is found, in terms of the radius of the contact area, radius of curvature of the probe tip, and the Poisson ratio and the modulus of elasticity of the probe tip and the sample. The environmental temperature and humidity are controlled to minimize heat transfer from the sample to the environment or to keep it constant and prevent any condensation on the sample. The heated block is shielded to reduce radiation heat transfer to the sample.

Heat, which leaves the probe tip, is expected to flow through a sequence of thermal resistances (probe-film interface, R_{p-f}/film, R_f/film-base interface, R_{f-b}/base, R_b/base-chuck interface, R_{b-c}), before it is eventually dumped in the temperature-controlled thermal chuck. The, comparator method works on the principle that the rate of heat transfer from the probe is controlled by the film's thermal resistance. We need to alert the users of this technique that, despite all of these precautions, there are still some factors that may distort the results. Some of these causes are discussed next.

1. A closer observation of the problem reveals that the area of contact between the probe tip and the sample surface is not the only path for heat transfer. The sample may also receive heat from the probe by radiation or conduction in the air. A simple calculation proves thermal radiation to be negligible, but heat conduction through the air can be significant, depending on conditions such as the contact pressure and radius of the contact area (see Table 3.3).

2. In measuring the high-thermal-conductivity films, the thermal resistance of the sample loses its controlling role, and other resistances, in particular, the contact resistances between the sample and the probe and base, R_{p-f} and R_{f-b}, take the role instead. The role of these two is particularly critical because they can be considerably different from one sample to another and with changes in the applied load. This condition violates the first requirements for a well-designed steady-state measurement technique [27]. Cheruparambil et al. [50] have observed a change in the comparator output by changing the applied pressure on a particular film, which was more pronounced for softer materials.

Table 3.3. The estimates of the thermal resistance, for different paths of heat transfer from the probe to the sample surface.

Radius of the contact area (μm)	Conduction in solid (K/W)	Conduction in air (K/W)	Radiation in air (K/W)
7	6.5×10^4	3.6×10^4	3.8×10^6
25.4	4.9×10^3	2.8×10^4	10^6

3. When the radius of the contact area of the probe tip reduces to the values of the same order of magnitude as the thickness of the film layer, the 2-D shape factor can become important [27]. This effect can, of course, be avoided if the samples in calibration and measurement have the same thickness.

3.5 Frequency-Domain Techniques

The frequency-domain techniques have been used extensively in the past for thermal characterization of the bulk materials and thin film layers. The frequency-domain techniques can be used to measure the directional thermal conductivities of the anisotropic film layers [51], [52], [53], [54] and the thermal boundary resistance between the layers. In these methods heat is usually generated by electrical Joule heating. It can be shown that when a metal line is subjected to harmonic Joule heating, the resulting temperature rise can be extracted from the third harmonic component of voltage oscillations across the line [55]. The optical heating has gained popularity due to its convenience and minimum microfabrication process requirements. However, due to the difficulty in determination of the absorbed absolute power from the optical beam, the extracted thermal property is restricted to measuring α, rather than both α and k. Thermocouples, thermistors, infrared sensors, or variation in thermal reflectance of a metallic layer can be used for thermometry. The measured temperature rise as a function of frequency can then be compared with the solution of the heat-conduction equation in the frequency domain, which allows extraction of the thermal transport properties of the film layers or substrates.

In this section, we will provide a general overview of the frequency-domain measurement techniques with different heating and thermometry schemes, different substrate (or films) configurations and geometries, and different locations of the heater or thermometer elements. Many of these techniques, such as the Ångström thermal wave (Sect. 3.5.1) and modified calorimetric techniques (Sect. 3.5.2), have been successfully implemented for thermal characterization of both low- and high-thermal-conductivity dielectrics [26], [27]. Whereas these techniques use periodic heating in one-dimensional structures or samples, more recently developed techniques can induce local heating in the geometry of a point source and thus achieve high spatial resolution in mapping sample thermal properties. One promising example is the study that applied local harmonic heating generated in a resistive element with a contact area dimension of around 30 nm along the surface of diamond layer [56]. The effects of contact topography, which governs thermal contact resistance between the heater and the layer, can easily be decoupled, because it results in frequency-independent contributions to the thermal signal. By directly comparing data with the data taken on the material with known thermal properties, Fiege et al. [56] spatially mapped the thermal conductivity of thick CVD-diamond film. Both the Ångström thermal wave and modified calorimetric techniques as well as steady-state [38] methods, can be used for the films on substrate if the film has a higher conductance than the substrate, which is discussed in Sect. 3.5.3. We also propose and examine new variations of the 3ω technique for the measurement of silicon-nitride (Si_3N_4) substrate (or layer) anisotropic thermal conductivity (Sect. 3.5.4) and nonuniform thermal conductivity across the AlN substrates (Sect. 3.5.5).

3.5.1 The Ångström Thermal Wave Technique

In the Ångström thermal wave technique [57], a periodic (electrical or optical) heat source with variable frequency is applied to a suspended thin film (or bar) in order to establish a one-dimensional thermal wave along the specimen (Fig. 3.15). The amplitude and phase of the temperature variations can be measured, either at the location of heating or along the suspended layer, to determine the thermal diffusivity α. Figure 3.16 shows the predicted reduced temperature rise at the heating location as a function of inversed root of frequency for a layer of thickness $30\,\mu m$ and thermal conductivity of $500\,Wm^{-1}K^{-1}$. Both in- and out-of-phase components can be used to extract the thermal properties. The plane source solution is well suited for the geometry of the suspended films (or bars). This method directly uses the result given in Eq. (3.4a) and Table 3.2:

$$\theta_{\text{plane}}(\omega) = \frac{Q_A}{A} \frac{1-i}{\sqrt{2}} k^{-1} \alpha^{\frac{1}{2}} \left[\omega^{-\frac{1}{2}} \right], \qquad (3.4a)$$

which is shown by solid lines. Obviously, the area A is twice the cross-sectional area of the film. There are two important issues in this measurement that require careful consideration, namely, surface (convective or radiative) heat loss from the suspended membrane and interaction of thermal waves with the

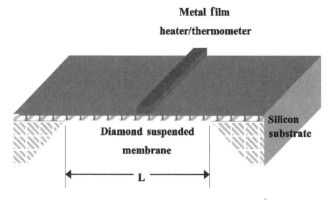

Fig. 3.15. Schematic of the measurement structure for the Ångström thermal wave technique.

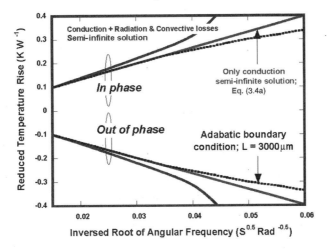

Fig. 3.16. Example of the 3ω measurement with plane heating source. The temperature signal deviates from predictions for the one-dimensional heating of a semi-infinite substrate, due to the surface losses, and the influence of the boundary condition in the actual suspended membrane geometry.

boundaries. Figure 3.16 also shows how the surface heat losses can alter the frequency response of the suspended membrane. The surface heat losses can be minimized by reducing the thermal penetration length, at sufficiently high frequencies [58], or be compensated for by using the geometric mean of diffusivity, obtained from phase and amplitude methods [59], [60]. In addition, Fig. 3.16 shows that the frequency response of the systems would deviate from the one-dimensional semi-infinite model, should the thermal waves reach the boundaries of the membrane. The interaction with the boundaries can also be limited by reducing the thermal penetration length in the suspended layer. While it is more appropriate to perform these measurements at high frequencies to limit the thermal penetration length, the 3ω signal would become increasingly small at these frequencies. As a result, it would make more sense to perform these measurements on sufficiently low frequencies and long samples while using the geometric mean of phase and amplitude data to eliminate surface loss effects.

3.5.2 The Modified Calorimetric Method

Another technique, which measures the amplitude of temperature oscillation as a function of distance from the heat source, was first introduced by Hatta [61] and then used by Yao [62] for thermal characterization of the 10-μ-thick free-standing AlAs/GaAs superlattice. In this method, which uses optical illumination as a heat source, the amplitude of the temperature oscillation, ΔT, is measured as a function of distance, x, between the edge of a movable mask and the thermocouple junction used to measure the film temperature

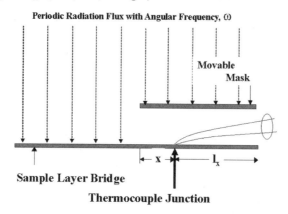

Fig. 3.17. Schematic of the measurement for the modified calorimetric technique.

(Fig. 3.17). The mask protects a part of the film layer from the laser light, and the exposed portion receives heat at a rate $Q = Q_A e^{i\omega t}$, where Q_A is the amplitude of absorbed heat flux. Assuming that heat loss from the surface is negligible, the solution for the temperature in the covered section of the plate, as a function of x, distance from the heat source, and frequency, is given as [59]

$$\theta(x,\omega) = \frac{Q_A}{2\omega C d} \exp[-\lambda x - i(\lambda x + \pi/2)], \qquad (3.15)$$

where, $\lambda = \sqrt{\omega/2\alpha}$ is reciprocal of the diffusion length. Plotting either $\ln|\theta(x,\omega)|$ or $\arg[\theta(x,\omega)]$ versus x, thermal diffusivity of the film can be extracted from the slope of the lines, which is equal to $\lambda(=\sqrt{\omega/2\alpha})$. If we know the specific heat, then the thermal conductivity can be calculated. Values of thermal diffusivity or thermal conductivity found from either amplitude or phase are the same, $\alpha = \alpha_a = \alpha_p$ and $\lambda = \lambda_a = \lambda_p$, if heat loss from the film surface is negligible. However, because heat loss from the surface is important, the amplitude and phase components of λ will be different ($\lambda_a > \lambda_p$). The correct values for λ and diffusivity will then be $\lambda = \sqrt{\lambda_a \lambda_p}$ and $\alpha = \sqrt{\alpha_a \alpha_p}$, respectively. As we see, the value of applied surface heat flux does not have any effect on the final results, and existence of the surface heat loss doesn't add any complexity to the problem.

In the recent round-robin measurements [32], different variations of heating and thermometry techniques were used for thermal property measurement of diamond samples. It appeared that the laboratories that used both phase and amplitude signals, employed scanned laser heating (stronger signals), and made corrections for boundary effects produced more reliable and consistent data for thermal diffusivity with very little variation from the mean value.

This method can also be used for the films on substrate, if the film has a higher conductance than the substrate, $dk_f \gg d_s k_s$, where d and k_f, and d_s and k_s are the thickness and conductivity of the film and substrate,

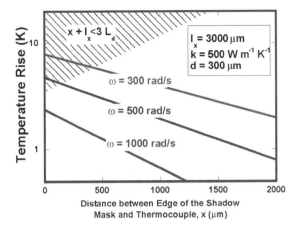

Fig. 3.18. The measured temperature rise as a function of thermocouple distance from the mask edge and frequency. Working in the hatched zone should be avoided for possible edge effects.

respectively [25]. Figure 3.18 shows the temperature rise measured by the thermocouple as a function of the heating frequency and distance between the thermocouple and edge of the moving mask. Possible reflection of the thermal waves from the edge (edge effects) for measurements in the hatched region can cause inaccuracy in the results.

3.5.3 The High-Thermal-Conductivity Films on the Low-Thermal-Conductivity Substrates

Both the Ångström thermal wave and modified calorimetric techniques can be used for films on substrate, if the film has a higher conductance than the substrate, $dk_f \gg d_o k_o$, where d and k_f, and d_o and k_o are the thickness and conductivity of the film and substrate, respectively [25], [63]. Figure 3.19 shows a schematic of the measurement structure that can be used to measure the thermal conductivity of the dielectric layers [63] or in general for thermal characterization of the layers with lateral thermal conductivity in the range of 10 to 2000 $Wm^{-1}K^{-1}$ and thickness of 0.5 to 10 μm. In this measurement, a metal bridge with variable width, deposited on the thin film, serves as both the heater and the thermometer. The thick silicon-dioxide layer under the sample acts as a thermal barrier. The lateral thermal conductivity of the thin film can be extracted by comparing the measured average temperature rise in the metal bridge with the analytical or numerical solution of the frequency-domain heat-diffusion equation.

The quality and microstructure of the CVD diamond is a strong function of the crystalline silicon substrate on which it is grown. However, the high thermal conductivity of the silicon substrate would significantly reduce the sensitivity of this measurement technique. As a result, it is crucial to grow

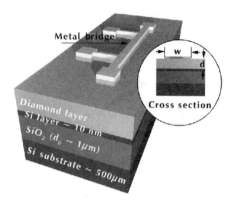

Fig. 3.19. The experimental structure for thermal property measurement of the ultrathin dielectrics and metallic layers.

high-quality CVD diamond on the ultrathin silicon-on-insulator (SOI) wafers to achieve high sensitivity in this measurement. The thickness of the silicon overlayer should be on the order of 10 nm or less to avoid any contribution of the lateral conduction in this layer to the measured thermal conductivity of the diamond film. This approach significantly reduces the need for removing the silicon substrate that requires relatively elaborate microfabrication process. In addition, the radiative and convective losses, which become increasingly important for very thin films, can be avoided altogether. The frequency-domain heat conduction analysis shows that, for layers deposited on $(d_0 =)$ 0.5-μm-thick SiO_2 film, one can achieve the required sensitivity for the measurement of lateral thermal conductivity of diamond layers at maximum power that the lock-in amplifier can dissipate in the metal bridge (Fig. 3.20).

3.5.4 Thermal Characterization of the Anisotropic Silicon-Nitride Substrates

As was discussed earlier, the orientation of β-Si_3N_4 grains induces a strong anisotropy in thermal conductivity of the silicon-nitride specimens [5]. It is clear that, in order to measure the directional thermal properties, one should confine the heat conduction and therefore the temperature gradient in a given direction. This is usually achieved by cutting three samples with perpendicular orientations (x, y, and z) and performing three independent measurements. However, the 3ω technique can be used to obtain the anisotropic thermal conductivity of β-Si_3N_4, using only one aluminum heater–thermometer bridge, which is deposited on the dielectric substrate. A schematic of the measurement structure and the thermal-conductivity coordinates are shown in Fig. 3.21.

Figure 3.22 shows sensitivity of the measurements to a 10% variation in the thermal conductivities k_x, k_y, and k_z for a given bridge length, at two

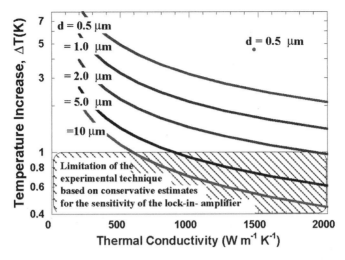

Fig. 3.20. Thickness restriction for the samples to ensure proper sensitivity for the temperature measurement.

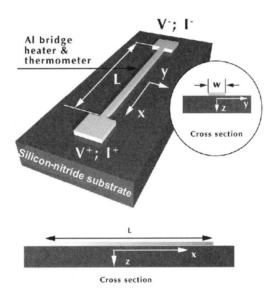

Fig. 3.21. Schematic of the structure, for 3ω measurements of the anisotropic thermal conductivities of the silicon nitride substrates ($\sim$500 μm).

different frequencies, $\omega = 1$ and 10,000 rad/s. It is clear that, depending on the measurement frequency and bridge length, the sensitivity of the measurement to thermal conductivity in a given direction varies substantially. For a long bridge, $L = 4000$ μm, and at high frequencies, $\omega = 10{,}000$ rad/s, the measurement is not sensitive to k_x, the thermal conductivity for the x-direction.

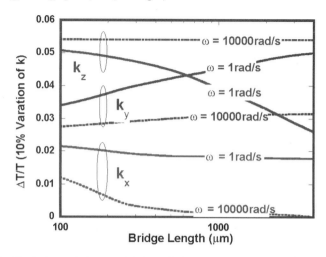

Fig. 3.22. Variation of the sensitivity of the measurements with the directional thermal conductivities, heater length, and frequency.

However, the measurement is sensitive to the conductivities in the y- and z-directions and, therefore, a combination of these two properties can be extracted. On the other hand, at low frequencies, the measurement becomes more sensitive to the variation in k_x, and sensitivity of the measurement to k_z is reduced so that a combination of k_x and k_y can be extracted. In the actual measurements, the thermal response of the substrate as a function of frequency can be measured using a lock-in amplifier. Subsequently, the data will be fitted to the three-dimensional heat conduction in the substrate over a range of frequencies from 1 to 10,000 rad/s from which thermal properties will be extracted. This procedure has been extensively used in the past for thermal characterization of different layers and substrates [53].

Figure 3.23 shows that, for a bridge of length $L = 4000\,\mu m$, the sensitivity of the measurements varies with the thickness of the substrates. For the substrates (layers) of thickness less than $200\,\mu m$, the measurement is more sensitive to k_x and k_y at low frequencies and to k_y and k_z at high frequencies. For the 20-μm-thick layer, the measurement is only sensitive to the k_y at high frequencies and to the combination of k_x and k_y at low frequencies.

3.5.5 Thermal Characterization of the AlN Substrates with Spatially Variable Thermal Conductivity

A strong gradient in the oxygen concentration of an AlN substrate was observed [20], which can translate to variations in thermal property across a given sample (Fig. 3.4). The 3ω technique can be used to probe different depths of a given substrate in order to extract its local thermal-transport properties. In this manner, one can avoid the complicated cutting process

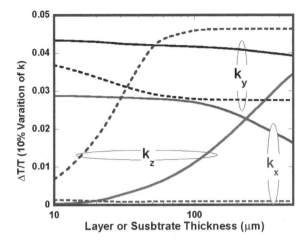

Fig. 3.23. Variation of the sensitivity of the measurements with thickness of the substrate (or layer) and frequency. Heater has a constant length, $L = 4000\,\mu$m. Dashed and solid lines correspond to the frequencies $\omega = 10{,}000$ and $1\,\text{rad/s}$, respectively.

and cumbersome repeated measurements. Because the heat penetration depth in the frequency-domain techniques is inversely proportional to the square root of the heating frequency, we can control the desired heated volume by controlling the heating frequency. Temperature oscillations in the heater are determined by the average thermal property for a given thermal penetration depth. By fitting the measured temperature rise in the heater to the predictions of the frequency-domain heat-conduction solution and assuming uniform thermal properties, we can obtain the effective thermal conductivity over the heated region. This provides valuable information for determining the local thermal conductivity at different depths away from the sample faces. For this purpose, a 2-D (finite volume) numerical model is developed for the 3ω measurements in the frequency domain. Thermal conductivity is considered to be a function of position in general. Figure 3.24 shows the simulated magnitudes of the temperature oscillation ΔT in the heater as a function of heating frequency from both top and bottom surfaces. The thermal conductivity in lateral direction, k_y, is assumed to be uniform along the depth and equal to $100\,\text{Wm}^{-1}\text{K}^{-1}$. The profile of the normal thermal conductivity, k_z, is given as $k_z(z) = 50 + 50e^{-1000z}\,\text{Wm}^{-1}\text{K}^{-1}$ from the top surface, as was shown in Fig. 3.4. The layer has a total thickness of 20 mm and extends infinitely in the lateral direction. The second surface of the layer is insulated and heated from the first surface. In the case of heating from the top surface, due to the fast decay of k_z in the near-surface region, deviation of the temperature rise from that resulting from a uniform k_z, starts at relative high heating frequency and becomes even larger as frequency decreases. This implies that more volume with lower thermal conductivity k_y is heated. However, the calculated

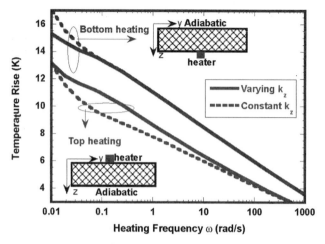

Fig. 3.24. Variation of the temperature rise with heating frequency for top and bottom surface probing.

temperature difference becomes smaller at very low frequencies, partly because, in this situation, the heat penetration depth exceeds the thickness of the layer and thus the boundary condition becomes important for determining the temperature rise. When heating the layer from its bottom surface, the calculated temperature rise is almost identical with that using a uniform value of $k_z = 50\,\mathrm{Wm^{-1}K^{-1}}$ over a large range of the heating frequency, until at some heating frequency the thermal waves reach the region of drastic variation in k_z. The lower the heating frequency, the larger the observed difference would be. Based on the temperature rise from two-side heating, we can somehow deduce the profile of the normal thermal conductivity k_z as a function of depth, or at least the varying tendency at different thicknesses.

3.5.6 The Mirage Technique

The mirage method is basically a modified version of the Ångström thermal wave technique that uses higher frequencies. During the measurement, a focused power laser is used to generate a localized heat source at the sample surface. Temporal modulation of the laser intensity results in oscillatory temperature distributions both in the body of the sample and on its surface. The induced temporal and spatial variations of the index of refraction of the air film in proximity to the heated surface are detected using a continuous probing laser aligned nearly parallel to the sample surface. The magnitude and phase of both the normal and transverse components of the refracted probe beam are monitored by a position-sensitive photodetector. The thermal diffusivity can be obtained by fitting the solution to a three-dimensional heat-diffusion equation for the relevant system (e.g., air, film, and substrate) to the experimental data [64], [65].

Two different schemes have been reported to detect the mirage signal for thermal-diffusivity measurement. In the transverse scheme, the probe laser beam travels in air skimming over the sample surface, heated by the pump beam, whose intensity is periodically modulated. This scheme has been used to measure samples with a wide range of thermal diffusivities from low, $10^{-7}\,\mathrm{m^2/s}$, such as for glass [66] and polymer [67] to high, $10^{-2}\,\mathrm{m^2/s}$, say for diamond [68]. The main concern in the transverse scheme is the skimming height of the probe beam from the sample surface. To avoid probe scattering on sample surface, the beam needs to travel some distance z away from the surface [69], [70]. For a low-thermal-diffusivity sample, the height z must be smaller than its thermal diffusion length $(\alpha/\omega)^{1/2}$, where α and ω are the sample thermal diffusivity and the heating modulation frequency, respectively. Otherwise, the measurement will seriously be affected by the heat diffusion in air. The surface reflection scheme is an alternative scheme to overcome the skimming height problem in which the probe beam is reflected on the sample surface close to the heating zone [68]. The disadvantage of this new scheme is that it cannot be applied to the nonreflecting samples, such as absorbing, rough, or transparent samples. One proposed solution to the problem has been deposition of a thin reflecting film on the sample surface. This, unfortunately, poses a new problem, which is the reflection of the pump beam.

Due to its noncontact configuration, the mirage technique does not require special sample preparations, especially for the standard transverse scheme. However, it involves more complicated experimental setup and data acquisition, than other thermal wave techniques.

3.6 Time-Domain Techniques

The time-domain techniques are based on application of heat to the sample's surface and monitoring its temperature changes with time. The source of heat is either electrical or optical. In the Joule heating method, the surface temperature rise is monitored during a relatively long ($\sim$100 μs) heating pulse. However, the laser heating method monitors the temperature relaxation after a very brief ($\sim$1 μs) heating pulse [16]. Due to its shorter time scale than the Joule heating method, the laser heating technique has a potential for investigating nonhomogeneities in the diamond thermal conductivity. Each of these techniques has advantages and disadvantages, which are discussed in the following sections. Goodson et al. [16] have used both Joule and laser heating methods to measure the normal thermal conductivity of the diamond layers. They were able to significantly improve the certainties of their diamond thermal-resistance measurements over previously reported works.

3.6.1 The Laser Heating Method

The laser heating method, which uses a laser for both heating and temperature measurement, is an example of a time-domain technique. In the most common experimental configuration, the film layer of interest is sandwiched between the substrate and a thick deposited metal film (Fig. 3.25). Heat flux normal to the layered structure is generated by the absorption of light, from a high-power laser, in the metal layer. The laser power is pulsed with a duration of ~10 ns. The temperature-induced changes of the metal surface reflectivity are then detected, using a low-power continuous-wave helium-neon laser. High thermal conductivity of the metal film will help fast absorption of the laser heat and its redistribution, causing a uniform temperature distribution. The metal film will also facilitate detection of changes in surface reflectivity.

This technique has proven to be extremely useful for characterization of optical and thermal properties of thin films, taking advantage of being non-contact and remote-sensing, and of the possibility for both high spatial and temporal resolution. The property extracted in this technique is the thermal resistance of the film layer. Stoner and Maris [71], using mode-locked lasers, measured the Kapitza resistance between the metal and a diamond film. Due to the high thermal conductivity, the effect of thermal conduction in diamond on the surface temperature rise is negligible. Goodson et al. [16] used the same technique to measure the total thermal resistance for conduction normal to the diamond layers on silicon substrate, to extract the thermal resistance of the diamond-silicon boundary.

An analytical solution of the transient thermal-diffusion problem in the layered material is necessary to identify the governing parameters and evaluate the thermal properties from the time-dependent thermal response. For the case of one-dimensional heat flow in a single-layer structure with the plane source geometry, the surface temperature, provided that the heating power

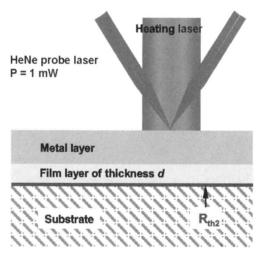

Fig. 3.25. Schematic of the laser heating method. Changes in surface temperature are measured by laser reflectance thermometry.

$Q(t)$ is an even function of time, $Q(t) = Q(-t)$, to satisfy $\varphi_0 = 0$, is given as

$$T(t) = \frac{1}{2\pi} k^{-1} \alpha^{\frac{1}{2}} \int_{-\infty}^{\infty} \left[\frac{Q(\omega)}{A} \frac{\cos(\omega t)}{\sqrt{2\omega}} \right] d\omega, \qquad (3.16a)$$

where

$$Q(\omega) = \int_{-\infty}^{\infty} Q(t) \cos(\omega t) dt. \qquad (3.16b)$$

Similarly, using the transmission-line theory [72], the frequency domain solution for a multilayer system, which accounts for the thermal-boundary resistances between adjacent layers, and the heating pulse shape factors can easily be obtained. For the one-dimensional transient heat diffusion in a metal-film-substrate system, the temperature rise on the top surface of the metal layer, under a Dirac function heating pulse at a Laplace frequency domain is given by

$$Z(s) = \frac{1}{e_1\sqrt{s}}$$

$$\times \frac{\left(\begin{array}{c} R_2 e_3 \sqrt{s} + 1 + e_{32} \tanh(\eta_2\sqrt{s}) + [1 + e_2\sqrt{s} R_2 \tanh(\eta_2\sqrt{s}) \\ + \tanh(\eta_2\sqrt{s})/e_{32}](e_{31}\tanh(\eta_1\sqrt{s}) + R_1 e_3\sqrt{s}) \end{array} \right)}{\left(\begin{array}{c} [1 + e_2\sqrt{s} R_2 \tanh(\eta_2\sqrt{s}) + \tanh(\eta_2\sqrt{s})/e_{32}] \\ \times [e_{31} + R_1 e_3\sqrt{s}\tanh(\eta_1\sqrt{s})] \\ + [R_2 e_3\sqrt{s} + 1 + e_{32}\tanh(\eta_2\sqrt{s})]\tanh(\eta_1\sqrt{s}) \end{array} \right)},$$

$$(3.17a)$$

where $e_i = k_i/\sqrt{\alpha_i}$, $i = 1, 2, 3$ are the thermal effusivity of the metal, film, and substrate; $\eta_i = d_i/\sqrt{\alpha_i}$, $i = 1, 2$ for the metal and thin film (the substrate is modeled as a semi-infinite medium); $e_{31} = e_3/e_1$, $e_{32} = e_3/e_2$; and finally, R_{th1} and R_{th2} are the thermal-boundary resistances at metal-film and film-substrate interfaces. Thus the time-domain surface temperature is given by the numerical inverse Laplace transform of $Z(s)$:

$$\theta_{Dirac}(t) = \frac{1}{2\pi i} \int_{\chi_0 - i\infty}^{\chi_0 + i\infty} Z(s) \exp(st) ds, \qquad (3.17b)$$

where the real-valued χ_0 is chosen to exceed the real part of any singularity of the integrand such as the origin. Once we know the temperature rise for a Dirac heating pulse, the temperature rise for heating pulses of any shape $\varphi(t)$ can be obtained by

$$\theta(t) = \int_0^t \varphi(t')\theta_{Dirac}(t - t')dt'. \qquad (3.17c)$$

Figure 3.26 shows the simulated surface temperature decay of the Au/Diamond/Si multilayer structure, under a 10-ns laser pulse. The thickness of the diamond layer is 5 μm, coated with a 1.5-μm Au layer and the silicon substrate is modeled as a semi-infinite medium. The temperature profile basically consists of three parts; first, a fast temperature rise to peak value from 0 to 10 ns, caused by the laser pulse heating; second, a fast temperature decay from 10 ns to 20 ns, due to the high thermal conductivity of the Au layer, as the thermal wave penetrates through it rapidly; and third, relatively slow temperature decay from 20 ns to 500 ns, which is determined by the thermal properties of the diamond layer and silicon substrate. The thermal resistance of the diamond layer is modeled either by the lumped thermal conductivity of diamond, which includes the combined thermal resistances of the film and the film-silicon interface, or by the summation of two thermal resistances caused by the film and its boundary with the silicon substrate. It is clear from the plots that the temperature decay curve based on the latter model falls out of the ±20% variation of the k band, in the lumped thermal-resistance model. This indicates that in thermal characterization of the high-thermal-conductivity films, the effect of thermal-boundary resistance at the film-substrate interface should be considered. At long time scales, when thermal-diffusion length in the silicon substrate is larger than the diamond-layer thickness, however, the shape of the temperature decay curve becomes sensitive to the combination of thermal conductivities of the diamond layer and silicon substrate, instead of the diamond-silicon boundary resistance.

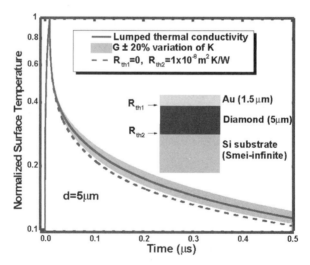

Fig. 3.26. Temperature decay on the surface of an Au/Diamond/Si system in the laser heating method. The 5-μm-thick diamond layer is coated with a 1.5-μm Au layer for the absorption of radiation energy. Solid line: temperature for lumped thermal conductivity model. Gray band: ±20% variation of the lumped thermal conductivity. Dashed line: Consider the thermal boundary resistance $R_{th2} = 1 \times 10^{-8}\,\mathrm{m^2 KW^{-1}}$ at the diamond/silicon interface ($R_{th1} = 0$ at Au/Diamond interface).

Graebner et al. [13] used the laser heating technique to measure the normal thermal conductivity of CVD diamond samples with thicknesses of 10–300 μm. Instead of the front-surface sensing, however, they used an infrared-detecting sensor to monitor the temperature rise at the rear face of the layer, as heat diffused through, provided that the laser pulse was short compared with the thermal diffusion time of the sample. The measured normal conductivity was at least 50% greater than the lateral conductivity. This is the direct result of the columnar microstructure of the CVD diamond.

3.6.2 The Joule Heating Method

The Joule heating method is another time-domain technique, which has been used to measure thermal properties of diamond films [16]. In a common test structure, the diamond film is deposited on a silicon wafer with a metal bridge deposited on its surface. The metal bridge has the dual responsibility of heating and thermometry. An electrical heating pulse with duration of 100 μs is applied to the metal bridge, and changes in the surface temperature are monitored by measuring the electrical resistance change along the bridge, which depends on temperature. By fitting the measured temperature to the transient three-dimensional thermal conduction equation, the thermal resistance for conduction normal to the diamond layer is calculated with $R_T = d/k_f$. The proper width for the bridge ($>$ film thickness) will ensure one-dimensional heat transfer in the direction normal to the film layer. Otherwise, lateral heat transfer in the diamond film will result in underprediction of the normal thermal resistance. The diamond film is assumed to have a homogeneous isotropic conductivity with a volume thermal resistance, which is larger than the boundary resistances at its interfaces with the metal bridge and the substrate. To estimate the thermal boundary resistance caused by the diamond-silicon interface, the measured temperature rise is compared with the temperature rise of a similar structure, excluding the diamond film. Using a doped diamond bridge, in the Joule heating method, there is also a potential for eliminating the metal diamond boundary resistance.

3.6.3 The Thermal Grating Technique

In the thermal grating technique, the thermal diffusivity is measured by the pulsed-laser excitation and detection of intensity change of the diffracted light, which is modulated by temperature variation on a sample surface. The sample is excited by two laser beams of spatial and longitudinal coherence, which are crossed inside the sample at an angle θ. As a result, the complex refraction index of the sample is periodically modulated with a period λ_G. The period of the induced grating-like structure can thus be varied by simply changing the crossing angle of the two pump beams. After the excitation, the heat is released by thermal diffusion from the sample surface in, creating a spatial sinusoidal temperature field on the sample surface with modulation amplitude ΔT, which decays as a function of time. As a consequence, the temperature dependence of the refractive index creates an optical (phase) grating in the sample. The temperature decay, then, can be detected by measuring the intensity of diffracted probe beam from the excited area on the sample surface. Because the thermal grating method probes the materials to a depth $\sim \lambda_G/\pi$, it seems to be suitable for the thermal property measurement of thin film materials, especially for those film-substrate structures. Kaeding et al. [35] demonstrated its application to the study of anisotropic lateral heat transport and thermal diffusivities of metal and diamond films. Marshall et al. [73] performed transient thermal grating experiments on a thin-film layer ($\sim 100\,$nm) of $YBa_2Cu_3O_7$ on MgO and $SrTiO_3$ substrates to measure the anisotropic thermal diffusivity and boundary resistance between the thin film and substrate. The lateral thermal diffusivity of the high-conductivity materials such as diamond was measured by adjusting the grating period λ_G to be much smaller than the light absorption length [34], [74], [75]. Compared to other time-domain techniques, this method involves a more complicated experimental apparatus and sample preparation for a highly reflecting surface.

3.7 Summary

This chapter described the available techniques and proposed new methods for measuring the highly anisotropic thermal properties of high-thermal-conductivity films and substrates. A variety of the steady-state and frequency- and time-domain techniques have been reported that are being used to measure the thermal properties of adamantine (diamond-like) materials. The choice of measurement technique for a particular experiment depends, obviously, on many factors, such as the required accuracy, the size and shape of the existing sample, and the availability of equipment and the skills to use it. Some of these techniques can measure the spatially variable properties, but others only measure the average values.

The steady-state measurements (e.g., heated bar) usually yield the highest level of accuracy and reliability, on the order of ± 5–10%, although one should carefully account for and minimize the convection and radiation losses from the specimen during the measurements. The accuracy of the measurements depends on the knowledge of the sample's precise dimensions. The proposed "film-on-substrate" technique can be used for routine thermal characterization of film layers with different thicknesses and, unlike the more conventional steady-state techniques, is entirely immune to errors due to the radiation and convection losses. The extension of the "classical comparator technique" to the thermal characterization of high-thermal-conductivity material seems promising, but the measurements require careful heat transfer analysis and a clear understanding of errors associated with the thermal-boundary resistance at the interfaces.

The Ångstrom thermal wave and the modified calorimetric methods require less effort in sample preparation but more careful consideration of the heater location and operational frequencies to minimize radiation losses and reduce the effect of the boundaries (e.g., the edge effects). Surface heat loss can be conveniently dealt with by combining the amplitude and phase solutions. By changing the shape of the heat source, either the thermal conductivity or diffusivity or a combination of the two properties can be extracted. These methods have been extensively used to measure both normal and lateral thermal properties. By proper choice of the bridge length and the frequency in the 3ω technique, the anisotropic thermal conductivity of the silicon-nitride substrates can be measured. The same technique can be used to characterize the aluminum-nitride substrates with spatially variable normal thermal conductivity. The mirage technique is even less dependent on sample dimensions with minimal need for specimen preparation and is ideally suited for films on substrates. The Ångstrom's thermal wave and the mirage techniques can yield relative uncertainties of ±5–10%, and ±5–15%, respectively [32]. These techniques require a high level of expertise and experience in the measurements and an in-depth understanding of the microscale heat-transfer processes.

The time-domain techniques, in particular the laser heating and thermal grating techniques, demand a relatively elaborate and expensive experimental apparatus. whereas the thermal reflectance technique (see Sect. 3.6.1) requires very little sample preparation, highly polished sample surfaces are needed for the thermal grating technique. The relative uncertainties are on the order of 15–20% [32], and at times even larger errors are expected.

Acknowledgment

The authors from Carnegie Mellon University would like to acknowledge support from the National Science Foundation (through grant number NSF-0103082 for Nanotechnology Interdisciplinary Research Team, NIRT) and the Data Storage System Center (DSSC) of Carnegie Mellon University.

References

[1] K. Watari, and S. L. Shinde, "High Thermal Conductivity Materials," *MRS Bull.* **26**, 440–1 (2001).

[2] N. Ichinose, and H. Kuwabara, *Nitride Ceramics*, Nikkan Kogyo-Shinbun, Tokyo (1998).

[3] H. Abe, T. Kanno, M. Kawai, and K. Suzuki, *Engineering Ceramics*, Gihodo-Shuppan, Tokyo (1984).

[4] K. Watari, "High Thermal Conductivity Non-oxide Ceramics," *J. Ceramic Soc. Japan,* **109**, S7–S16 (2001).

[5] K. Hirao, K. Watari, H. Hayashi, and M. Kitayama, "High Thermal Conductivity Silicon Nitride Ceramic," *MRS Bull.* **26**, 451–5 (2001).

[6] K. E. Goodson, "Impact of CVD Diamond Layers on the Thermal Engineering of Electronic Systems," *Ann. Rev. Heat Trans.* **6**, 323–53 (1995).

[7] J. S. Goela, N. E. Brese, M. A. Pickering, and J. E. Graebner, "Chemical Vapor Deposited Materials for High Thermal Conductivity Applications," *MRS Bull.* **26**, 458–63 (2001).

[8] A. Eucken, "The Heat Conductivity of Certain Crystals at Low Temperatures," *Phys. Z* **12**, 1005–8 (1911).

[9] R. Berman, F. E. Simon, and J. Wilks, *Nature*, **168**, 277 (1951).

[10] J. E. Graebner, "Thermal Conductivity of CVD Diamond: Techniques and Results," *Diamonds Film and Technology*, **3**, 77–130 [1993].

[11] M. N. Touzelbaev, and K. E. Goodson, "Applications of Micron-Scale Passive Diamond Layers for the Integrated Circuits and Microelectromechanical Systems Industries," *Diamond and Related Materials* **7**, 1–14 (1998).

[12] J. E. Graebner, S. Jin, G. W. Kammlott, J. A. Herb, and C. F. Gardinier, "Unusually High Thermal Conductivity in Diamond Films," *Appl. Phys. Lett.* **60**, 1576–8 (1992).

[13] J. E. Graebner, S. Jin, G. W. Kammlott, B. Bacon, and L. Seibles, "Anisotropic Thermal Conductivity in Chemical Vapor Deposition Diamond," *J. Appl. Phys.* **71**, 5353–6 (1992).

[14] J. E. Graebner, S. Jin, G. W. Kammlott, J. A. Herb, and C. F. Gardinier, "Large Anisotropic Thermal Conductivity in Synthetic Diamond Films," *Nature* **359**, 401–3 (1992).

[15] R. Csencsits C. D. Zuiker, D. M. Gruen, and A.R. Krauss, "Grain Boundaries and Grain Size Distributions in Nanocrystalline Diamond Films Derived from Fullerene Precursors," *Diffusion and Defect Data*-Part B (Solid State Phenomena), Vol. 51–2, 261–9 (1996).

[16] K. E. Goodson O. W. Kaeding, M. Rosler, and R. Zachai, "Experimental Investigation of Thermal Conduction Normal to Diamond Silicon Boundaries," *J. Appl. Phys.* **77**, 1385–92 (1995).

[17] K. Watari, Y. Seki, and K. Ishizaki, "Temperature Dependence of Thermal Coefficients for HIPped Silicon Nitride," *Journal of Ceramic Society of Japan* **97**, 174–81 (1989).

[18] K. Watari, M. E. Brito, M. Toriyama, K. Ishizaki, S. Cao, and K. Mori, "Thermal Conductivity of Y_2O_3-doped Si_3N_4 Ceramic at 4 to 1000 K," *J. Mat. Sci. Lett.* **18**, 865–79 (1999).

[19] G. A. Slack, "Nonmetallic Crystals with High Thermal Conductivity," *J. Phys. and Chem. of Solids* **34**, 321–35 (1973).

[20] H. Buhr, G. Muller, H. Wiggers, F. Aldinger, P. Foly, and A. Roosen, "Phase Composition, Oxygen Content, and Thermal Conductivity of Aluminum Nitride (yttria) Ceramics," *J. Amer. Cer. Soc.* **74**, 718–22 (1991).

[21] C. Pelissonnier, L. Pottier, D. Fournier, and A. Thorel, "High-Resolution Thermal Diffusivity Measurements in Aluminum Nitride. Effect on Grain Boundaries and Intergranular Phases," *Mat. Sci. Forum* **3**, 207–9 (1996).

[22] J. Kölzer, E. Oesterschulze, and G. Deboy, "Thermal imaging and measurement techniques for electronic materials and devices," *Microelectronic Engineering*, **31**, 251–270 (1996).

[23] M. Asheghi, Y. K. Leung, S. S. Wong, and K. E. Goodson, "Phonon-Boundary Scattering in Thin Silicon Layers," *Appl. Phys. Lett.* **71**, 1798–1800 (1997).

[24] K. E. Goodson and Y. S. Ju, "Heat Conduction in Novel Electronic Films," *Ann. Rev. Mat. Sci.* **29**, 261–93 (1999).

[25] K. E. Goodson and M. I. Flik, "Solid Layer Thermal-Conductivity Measurement Techniques," *Appl. Mech. Rev.* **47**, 101–12 (1994).

[26] D. G. Cahill, "Heat Transport in Dielectric Thin Films and at Solid-Solid Interfaces," *Microscale Energy Transport* (C. L. Tien, A. Majumdar, and F. M. Gerner, ed.), 95–117 (1998).

[27] M. N. Touzelbaev, and K. E. Goodson, "Impact of Experimental Timescale and Geometry on Thin Film Thermal Property Measurements," *Proc. 14th Symp. on Thermophysical Properties*, June 25–30, 2000, Boulder, Color (2000).

[28] K. E. Goodson, "Thermal Conduction in Electronic Microstructures," in the *Thermal Engineering Handbook*, F. Kreith, ed., CRC Press, Boca Raton, Florida, pp. 4.458–4.504 (2000).

[29] K. Plamann, D. Fournier, B. C. Forget, and A. C. Boccara, "Microscopic Measurements of the Local Heat Conduction in Polycrystalline Diamond Films," *Diamond and Related Mat.* **5**, 699–705 (1996).

[30] D. Fournier and K. Plamann, "Thermal Measurements on Diamond and Related Materials," *Diamond and Related Mat.* **4**, 809–19 (1995).

[31] J. E. Graebner, "Thermal Measurement Techniques," *Handbook of Industrial Diamonds and Diamond Films* (M. A. Prelas, G. Popovici, and L. K. Bigelow, eds.), Marcel Dekker, New York, 193–225 (1997).

[32] J. E. Graebner, H. Altmann, N. M. Balzaretti, R. Campbell, H.-B. Chae, A. Degiovanni, R. Enck, A. Feldman, D. Fournier, J. Fricke, J. S. Goela, K. J. Gray, Y. Q. Gu, I. Hatta, T. M. Hartnett, R. E. Imhof, R. Kato, P. Koidl, P. K. Kuo, T.-K. Lee, D. Maillet, B. Remy, J. P. Roger, D.-J. Seong, R. P. Tye, H. Verhoeven, E. Worner, J. E. Yehoda, R. Zachai, and B. Zhang, "Report on a Second Round Robin Measurement of the Thermal Conductivity of CVD Diamond," *Diamond and Related Materials* **7**, 1589–604 (1998).

[33] W. S. Capinski, and H. J. Maris, "Improved Apparatus for Picosecond Pump-and-Probe Optical Measurements," *Rev. Sci. Instruments* **67**, 2720–6 (1996).

[34] O. W. Kaeding, E. Matthias, R. Zachai, H.-J. Füßer, and P. Münzinger, "Thermal Diffusivities of Thin Diamond Films on Silicon," *Diamond and Related Materials* **2**, 1185–90 (1993).

[35] O. W. Kaeding, H. Skurk, A. A. Maznev, and E. Matthias, "Transient Thermal Gratings at Surfaces for Thermal Characterization of Bulk Materials and Thin Films," *Appl. Phys. A*, **61**, 253–61 (1995).

[36] H. Shibata, H. Ohta, and Y. Waseda, "New Laser-Flash Method for Measuring Thermal Diffusivity of Isotropic and Anisotropic Thin Films," *JIM Mat. Trans.* **32**, 837–44 (1991).

[37] H. S. Carslaw, and J. C. Jaeger, *Conduction of Heat in Solids*, second edition, Oxford Science Publications, Oxford (1959).

[38] M. Asheghi, M. N. Touzelbaev, K. E. Goodson, Y. K. Leung, and S. S. Wong, "Temperature Dependent Thermal Conductivity of Single Crystal Silicon Layers in SOI Substrates," *J. Heat Trans.* **120**, 30–6 (1998).

[39] D. T. Morelli, C. Uher, and C. J. Robinson, "Transmission of Phonons Through Grain Boundaries in Diamond Films," *Appl. Phys. Lett.* **62**, 1085–7 (1993).

[40] J. E. Graebner, M. E. Reiss, and L. Seibles, "Phonon Scattering in Chemical Vapor Deposited Diamond," *Phys. Rev. B* **50**, 3702–13 (1994).

[41] E. Worner, C. Wild, W. Muller-Sebert, M. Funer, M. Jehle, P. Koidl, "Thermal Conductivity of CVD Diamond Films: High-Precision, Temperature-Resolved Measurements," *Diamond and Related Materials* **5**, 688–92 (1996).

[42] G. Pompe, and K. Schmidt, "Vapor Deposited Lead Films and Their Transport Characteristics at Low Temperatures," *Phys. Stat. Sol.* **A31**, 37–46 (1975).

[43] M. Asheghi, K. Kurabayashi, R. Kasnavi, and K. E. Goodson, "Thermal Conduction in Doped Single-Crystal Silicon Films," *J. Appl. Phys.* **91**, 5079–88 (2002).

[44] J. E. Graebner, J. A. Mucha, L. Seibles, and G. W. Kammlott, "The Thermal Conductivity of Chemical Vapor Deposited Diamond Films on Silicon," *J. Appl. Phys.* **71**, 3143–6 (1992).

[45] E. Jansen, O. Dorsch, E. Obermeier, and W. Kulisch, "Thermal Conductivity Measurements on Diamond Films based on Micromechanical Devices," *Diamond and Related Materials* **5**, 644–8 (1996).

[46] E. Jansen, and E. Obermeier, "Thermal Conductivity Measurements on Thin Films based on Micromechanical Devices," *J. Micromech. and Microeng.* **6**, 118–21 (1996).

[47] E. Jansen, and E. Obermeier, "Measurements of the Thermal Conductivity of CVD Diamond Films Using Micromechanical Devices," *Phys. Stat. Sol. (a)* **154**, 395–402 (1996).

[48] R. W. Powell, "Experiments Using a Simple Thermal Comparator for Measurement of Thermal Conductivity, Surface Roughness and Thickness of Foils or of Surface Deposits," *J. Sci. Instruments* **34**, 485–92 (1957).

[49] J. C. Lambropoulos, M. R. Jolly, C. A. Amsden, S. E. Gilman, M. J. Sinicropi, D. Diakomihalis, and S. D. Jacobs, "Thermal Conductivity of Dielectric Thin Films," *J. Appl. Phys.* **66**, 4230–42 (1989).

[50] K. R. Cheruparambil, B. Farouk, J. E. Yehoda, and N. A. Macken, "Thermal Conductivity Measurement of CVD Diamond Films Using a Modified Thermal Comparator Method," *J. Heat Trans.* **122**, 808–16 (2000).

[51] A. R. Kumar, D.-A. Achimov, T. Zeng, and G. Chen, "Thermal Conductivity of Nanochanneled Alumina," *ASME HTD* **366**–**2**, 393–7 (2000).

[52] T. Borca Tasciuc, W. L. Liu, J. L. Liu, T. F. Zeng, D. W. Song, C. D. Moore, G. Chen, K. L. Wang, M. S. Goorsky, T. Radetic, R. Gronsky, T. Koga, and M. S. Dresselhause, "Thermal Conductivity of Symmetrically Strained Si/Ge Superlattices," *Superlattices Microstructure* **28**, 199–206 (2000).

[53] K. Kurabayashi, M. Asheghi, M. Touzelbaev, and K. E. Goodson, "Measurement of the Thermal Conductivity Anisotropy in Polyimide Films," *IEEE J. Microelectromechanical Systems* **8**, 180–91 (1999).

[54] Y. S. Ju, K. Kurabayashi, and K. E. Goodson, "Thermal Characterization of Anisotropic Thin Dielectric Films Using Harmonic Joule Heating," *Thin Solid Films* **339**, 160–4 (1999).

[55] D. G. Cahill, "Thermal Conductivity Measurement from 30–750 K: The 3ω Method," *Rev. Sci. Instruments* **61**, 802–8 (1990).

[56] G. B. M. Fiege, A. Altes, A. Heiderhoff, and L. J. Balk, "Quantitative Thermal Conductivity Measurements with Nanometer Resolution," *J. Phys. D* **32**, L13–L17 (1999).

[57] A. J. Ångström, "Neue Methode das Wärmeleitungsvermögen der Körper zu Bestimmen," *Ann. Phys. Chem.* **114**, 513–30 (1861).

[58] M. B. Salamon, P. R. Garnier, B. Golding, and E. Buehler, "Simultaneous Measurement of the Thermal Diffusivity and Specific Heat Near Phase Transitions," *J. Phys. Chem. of Solids* **35**, 851–9 (1974).

[59] Y. Gu, X. Tang, Y. Xu, and I. Hatta, "Ingenious Method for Eliminating Effects of Heat Loss in Measurements of Thermal Diffusivity by AC Calorimetric Method," *Japanese J. Appl. Phys.* **32**, L1365–7 (1993).

[60] Y. Gu, D. Zhu, L. Zhu, and J. Ye, "Thermal Diffusivity Measurement of Thin Films by the Periodic Heat Flow Method with Laser Heating," *High Temperatures—High Pressures* **25**, 553–9 (1993).

[61] I. Hatta, "Thermal Diffusivity Measurement of Thin Films by Means of an AC Calorimetric Method," *Rev. Sci. Instruments* **56**, 1643–7 (1985).

[62] T. Yao, "Thermal Properties of AlAs/GaAs Superlattices," *Appl. Phys. Lett.* **51**, 1798–1800 (1987).

[63] Y. S. Ju, and K. E. Goodson, "Process-Dependent Thermal Transport Properties of Silicon-Dioxide Films Deposited Using Low-Pressure Chemical Vapor Deposition," *J. Appl. Phys.* **85**, 7130–4 (1999).

[64] A. C. Boccara, D. Fournier, and J. Badoz, "Thermooptical Spectroscopy: Detection by the Mirage effect," *Appl. Phys. Lett.*, **36**, 130–2.

[65] A. C. Boccara, D. Fournier, W. Jackson, and N. M. Amer, "Sensitive Photothermal Deflection Technique for Measuring Absorption in Optically Thin Media," *Opt. Lett.* **5**, 377–9 (1980).

[66] M. Bertolotti, G. Liakhou, R. Li Voti, F. Michelotti, and C. Sibilia, "Method for Thermal Diffusivity Measurements Based on Photothermal Deflection," *J. Appl. Phys.* **74**, 7078–84 (1993).

[67] J. Rantala, W. Lanhua, P. K. Kuo, J. Jaarinen, M. Luukkala, and R. L. Thomas, "Determination of Thermal Diffusivity of Low-Diffusivity Materials Using the Mirage Method with Multiparameter Fitting," *J. Appl. Phys.* **73**, 2714–23 (1993).

[68] T. R. Anthony, W. F. Banholzer, J. F. Fleischer, W. Lanhua, P. K. Kuo, R. L. Thomas, and R. W. Pryor, "Thermal Diffusivity of Isotopically Enriched Sup 12/C Diamond," *Phys. Rev. B* **42**, 1104–11 (1990).

[69] Z. L. Wu, M. Thomsen, P. K. Kuo, Y. S. Lu, C. Stolz, and M. Kozlowski, "Overview of Photothermal Characterization of Optical Thin Film Coatings," *Proc. SPIE—The International Society for Optical Engineering*, **2714**, 465–81 (1986).

[70] A. Salazar, A. Sanchez-Lavega, and J. Fernandez, "Photothermal Detection and Characterization of a Horizontal Buried Slab by the Mirage Technique," *J. Appl. Phys.* **70**, 3031–7 (1991).

[71] R. J. Stoner, and H. J. Maris, "Kapitza Conductance and Heat Flow Between Solids at Temperatures from 50 to 300 K," *Phys. Rev. B* **48**, 16373–87 (1993).

[72] P. Hui and H. S. Tan, "A Transmission-Line Theory for Heat Conduction in Multilayer Thin Films," *IEEE Trans. Components, Packaging, and Manufacturing Technology*, Part B **17**, 426–34 (1994).

[73] C. D. Marshall, A. Tokmakoff, I. M. Fishman, and M. D. Fayer, "Interface Selective Transient Grating Spectroscopy: Theory and Applications to Thermal Flow and Acoustic Propagation in Superconducting Thin Films," *Proc. SPIE—The International Society for Optical Engineering* **1861**, 302–13 (1993).

[74] J. E. Graebner, "Measurement of Thermal Diffusivity by Optical Excitation and Infrared Detection of a Transient Thermal Grating," Review of Scientific Instruments, **66**, 3903–06 (1995).

[75] Y. Taguchi, and Y. Nagasaka, "Thermal Diffusivity Measurement of High-Conductivity Materials by Dynamic Grating Radiometry," *Int. J. Thermophys.* **22**, 289–99 (2001).

4

Thermal Wave Probing of High-Conductivity Heterogeneous Materials

Danièle Fournier

Thermal wave probing has turned out to be a very interesting way to study thermal diffusivity of materials at various scales without much metallographic preparation. Understanding thermal conductivity and microstructure of heterogeneous materials is more difficult than understanding those of single crystal because they consist of grains and grain boundaries. The knowledge of thermal parameters at various spatial scales is necessary because macroscopic thermal properties of these materials are strongly dependent on the microscopic properties of grains and grain boundaries. We describe thermal wave techniques for characterizing thermal properties of materials with heterogeneous microstructure.

In the first part, photothermal experimental setups are described: detection of thermal waves can indeed be done with different experimental setups depending on the physical parameters under investigation (refractive index, infrared emissivity, acoustic waves, local deformation). Then we show which pertinent parameters can be extracted from this kind of experiment. In the second part, we illustrate the ability of our mirage setup to study materials at millimeter scales; judicious choices of frequency modulation and experimental conditions geometry allow us to measure thermal properties of very thin samples, such as 1-micrometer-thick layers of synthetic diamond, or to evaluate thermal resistance interfaces. In the third part, our photoreflectance microscope is described in detail, and we underline all the abilities of this unique experimental setup to determine thermal diffusivity of individual grain in a ceramic sample such as AlN ceramic. The same setup is also able to evaluate thermal resistance at the grain boundaries of heterogeneous samples. We conclude with a brief comparison of the abilities of all these setups with more classical methods.

4.1 Introduction

In recent years, high-thermal-conductivity materials such as ceramics and films have been developed. Understanding thermal conductivity and microstructure of these materials is more difficult than understanding single crystal, because they consist of grains and grain boundaries. Because macroscopic thermal properties of these materials are strongly dependent on the microscopic properties of grains and grain boundaries, knowledge of thermal parameters at various spatial scales is necessary. Thermal wave probing has turned out to be a very interesting way to study thermal diffusivity of materials at various scales without much metallographic preparation. In this review, we describe this technique for characterizing thermal properties of materials with heterogeneous microstructure.

4.2 Thermal Parameter Determination with a Photothermal Experiment

4.2.1 Photothermal Experiment Principle

A photothermal experiment is the "combination" of the illumination of the sample with a pulsed or modulated pump beam and the detection of the surface or volume temperature variations related to the transformation of this radiant energy into heat. This measurement leads to the determination of the thermal properties of the sample.

   This detection can be done with different experimental setups depending on the physical parameters, which are affected by the temperature variation: refractive index, infrared emissivity, acoustic waves, or local deformation. In this chapter, we use mainly two kinds of optical detection: the mirage effect for 1D or 3D experiments and the photoreflectance for 3D configuration setups, both of which allow detection of the surface or the bulk temperature variation with a very good sensitivity.

4.2.2 Plane and Spherical Thermal Waves

When the heat source is intensity modulated, the diffusion of the heat can be described with the concept of a "thermal wave." To explain the concept of a thermal wave, let us solve the heat diffusion equation in a semi-infinite medium with a plane modulated heat source (absorbed modulated light flux) located at $x = 0$. The solution is the following:

$$T(x,t) = K \exp(-x/\mu) \exp(j[\omega t - \pi/4 - x/\mu]), \qquad (4.1)$$

where $T(x,t)$ is the temperature at a distance x from the sample surface (see Fig. 4.1), μ is the thermal diffusion length that characterizes the heat propagation in the medium and can be controlled by the modulation frequency ($f = \omega/2\pi$) of the heat source; μ is related to κ (thermal conductivity), ρ

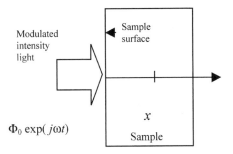

Fig. 4.1. Creation of the modulated heat source at the sample surface.

(density), C (specific heat), and f by the following expression:

$$\mu = (\kappa/2\pi f \rho C)^{1/2} = (D/2\pi f)^{1/2},$$

where D is the thermal diffusivity and $K = \Phi_0/\sqrt{w}\sqrt{\kappa\rho C}$, where Φ_0 is the light power (W/m^2) and $\sqrt{\kappa\rho C} = e$ is the effusivity of the sample.

Figure 4.2 clearly demonstrates that the heat propagation is strongly damped if x is larger than $2\pi\mu$. Moreover, expression (4.1) could describe the propagation of an optical wave for which the real part and the imaginary part of the refractive index would be equal. A thermal wave, like an optical wave, can reflect, diffract, and propagate. Note that Eq. (4.1) clearly indicates that measuring the phase lag of the complex temperature allows both the measurement of the thermal diffusion length and the localization of heat sources in the sample.

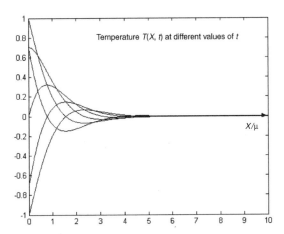

Fig. 4.2. Temperature inside the sample at various times ($\omega t = 0, \pi/4, \pi/2, \pi \ldots$) versus x/μ.

To carry out a thermal diffusivity measurement with our setups, a focused heating spot has to be used. Therefore, to interpret the results, the heat diffusion equation has to be solved in a spherical $T(R,t)$ or cylindrical geometry $T(r,z,t)$. Fourier and Hankel transforms then allow easy calculations in the case of modulated heat sources. Let us simply outline that in the case of a heat point source, the result is given by:

$$T(R,t) = T_0 \exp\left(-\frac{R}{\mu}\right) \exp\left(j\left(wt - \frac{R}{\mu}\right)\right). \tag{4.2}$$

Thus, for the heat diffusion equation in a homogeneous material, we have a **spherical thermal wave**. Let us underline that the thermal diffusion length can also be measured with the phase lag versus the distance to the heat source point.

4.2.3 Thermal Conductivity, Thermal Diffusivity, and Thermal Effusivity

In order to determine the thermal behavior of a sample or component we have to determine three parameters, the thermal conductivity κ, the thermal diffusivity D, and the thermal effusivity e. The first one, κ, governs the transfer of energy within a material. The rate of heat transfer depends on the temperature gradient and the thermal conductivity of the material. The second one, D, characterizes heat conduction in a nonstationary regime, and the last one characterizes the thermal contact of two samples.

In a modulated photothermal experiment, the sample is heated with a modulated heat source. We saw in the previous section that, whatever the geometry of the experiment (plan or point heat sources), a thermal wave propagates in the sample with a characteristic length μ. This length can easily be extracted from a photothermal experiment by measuring the phase lag of the local temperature versus the distance from the heat source. The thermal diffusivity, then, is the pertinent parameter of a photothermal experiment.

The thermal conductivity and the thermal effusivity are more difficult to obtain directly because they need absolute measurements.

4.2.4 Thermal Waves and Photothermal Setups

The measurement of the thermal diffusion length can be done easily if the phase and the amplitude of the surface temperature are realized. In our laboratory we have developed three setups able to do such measurement: the mirage setup, the photothermal microscope, and the infrared microscope. In the first, we take into account the temperature gradient in the adjacent fluid above the sample, associated to the surface temperature variations. The deflection of a laser beam propagating in this refractive index gradient is a measurement of the surface temperature and then of the thermal diffusivity length. In the photothermal microscope, a probe beam is reflected on the surface of the sample. The measurement of the variation of this reflected light allows the determination of the surface temperature variation. Finally, it is also possible to measure the variation of the emissivity of the surface sample with an infrared detector to detect the surface temperature.

In the three cases, we have to scan the heated area to get information on both amplitude and phase. The scan is achieved by moving the mirage cell, the probe beam, or the infrared detector. The modulation frequency of the beam intensity is achieved with either a mechanical chopper or an acousto-optic modulator. The modulation frequency range of all experiments extends from 10 Hz to 1 MHz. The thermal diffusion length for a given sample can be reduced by 300 when the modulation frequency goes from 10 Hz to 1 MHz.

To understand the behavior of complex materials such as ceramic, film, and composite, it is important to know not only the physical parameters at a macroscopic scale associated to the average behavior but also the thermal parameters of each constituent at microscopic scale. Photothermal experiments that are able to determine the thermal diffusivity at different spatial scales from the micrometer to the millimeter by only varying the modulation frequency are good solutions for the investigation of complex materials.

4.2.5 Analysis of the Experimental Data

4.2.5.1 Bulk Sample. When the sample is illuminated with a laser beam, we have to take into account the Gaussian repartition of the flux and convolute the heat point source solution with the Gaussian repartition. Moreover, the detection is done with a finite size detector, so we also have to take into account the detector size.

Figure 4.3 shows the surface temperature in amplitude (log scale) and in phase (degrees) versus the distance of the laser beam in the case of a 100-μm-diameter Gaussian excitation at 1000 Hz. Far from the center, where the shape of the curve is strongly affected by both the width of the pump beam and the detector size, the phase curves are linear versus the distance to the heat source, and the slope $(-1/\mu)$ is inversely proportional to μ. We can read directly on the figure the thermal diffusion length for one radian variation Δr equals μ (μ gold $= 200 \, \mu$m and μ diamond $= 560 \, \mu$m). Figure 4.4 shows the behavior of the thermal propagation at 1 MHz. As expected, the thermal diffusion length is inversely proportional to the square root of the frequency, and μ is 6.2 μm for gold and 17 μm for diamond. Whatever the frequency, the measurements of the slope give directly the thermal diffusion length.

4.2.5.2 Influence of the Sample Thickness. Formula (4.2) corresponds to the heat propagation in an infinite sample. For a given sample thickness we have to be careful with the choice of the modulation frequency. If the sample thickness is smaller or of the same order of magnitude as the thermal diffusion length, the slope does not give directly the expected thermal diffusion length. We have to solve the heat diffusion equation, taking into account heat diffusion

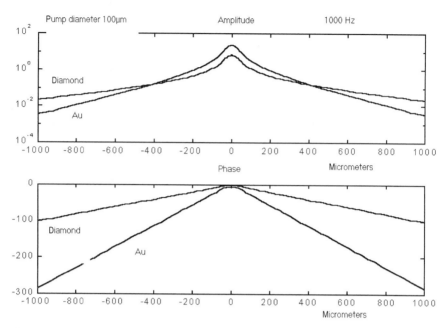

Fig. 4.3. Surface temperature of gold and diamond samples illuminated with 100-μm-diameter pump beam (1000 Hz); μ is obtained for a phase lag of 57°.

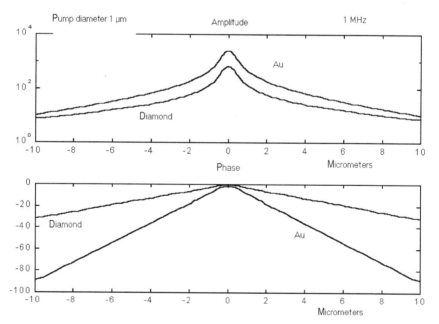

Fig. 4.4. Surface temperature of gold and diamond samples illuminated with 1-μm-diameter pump beam (1 MHz); μ is obtained for a phase lag of 57°.

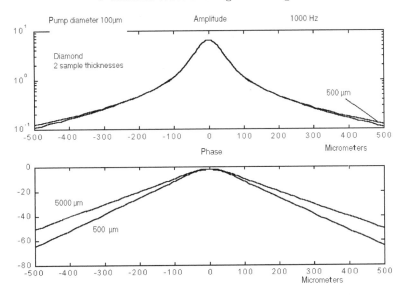

Fig. 4.5. Influence of the sample thickness at a given modulation frequency $(1000 \, \text{Hz}; \, D = 10 \, \text{cm}^2/\text{s}; \mu = 564 \, \mu\text{m})$.

in three media: air, sample, and air (or backing). Figure 4.5 illustrates the case of a finite sample thickness; the slope is not an actual measurement of the thermal diffusion length if the thermal wave reflects at the bottom of the sample (case of the 500-μm-diamond thick).

4.2.5.3 Absorbing and Reflected Layer Deposited on Top of the Sample. We generally use an argon laser to create the heat source at the sample surface. If the sample is transparent or scatters the laser beam, we have to deposit an absorbing layer. For the photothermal microscope setup, a 100-nm gold layer is usually deposited on top of the sample; for the infrared microscope a 100-nm Au/Pd layer is used. We have to know whether this layer has to be taken into account when the thermal diffusivity of the substrate is measured. To know the temperature distribution in the whole assembly we have to solve the heat diffusion equation in all the media (fluid, coating, sample, and backing). In the case of a layered sample the main parameter is the effusivity of each layer. At each interface, the incident thermal amplitude (A_0) is transmitted (A_1) and reflected (B_0) according to the thermal properties of the media with the thermal reflexion coefficient $B_0/A_0 = e_0 - e_1/e_0 + e_1$ where the analog of the refractive index is the thermal effusivity e. To illustrate, consider a gold layer deposited on three substrates. Table 4.1 gives the thermal reflection coefficients for the three substrates.

Figure 4.6 illustrates the influence of a 100-nm gold layer deposited on three samples. In this case, the thermal diffusion length in the layer is larger than the thickness (thermally thin case). Following the thermal mismatch

Table 4.1.

	Gold	Platinum	Silicon	Diamond
Effusivity ($Jm^{-2}K^{-1}s^{-1/2}$)	2.8×10^4	1.44×10^4	1.58×10^4	3.79×10^4
Thermal reflection coefficient/gold	0	0.32	0.27	−0.15

at the interface, this layer has to be taken into account ($R = 0.32$) or not ($R = -0.15$ and $R = 0.28$). When the sample is a better or equal conductor than the layer, we can ignore it in our analysis. But if the substrate thermal conductivity is lower than the gold thermal conductivity (in the case of platinum), we have to use a multilayered model to determine the thermal diffusivity of the substrate. In conclusion, it seems wise to achieve a complete calculation in all cases.

4.2.5.4 Multilayered Samples. When the sample under investigation is multilayered, it is possible to extract the thermal diffusivity of the layer if the thickness of the layer and the thermal properties of the substrate are known.

Moreover, if the thermal properties of the layer and the substrate are unknown, in some cases we can develop the following strategy.

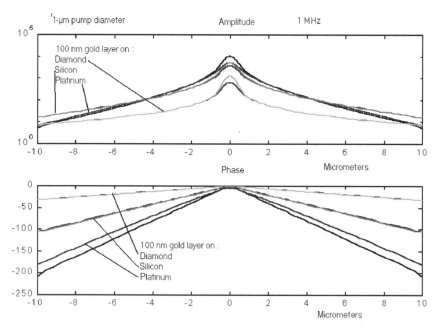

Fig. 4.6. Influence of a 100-nm gold coating deposited on different samples: diamond, silicon, and platinum at 1 MHz. In two cases (diamond and silicon), the slopes give directly the values of the thermal diffusion length.

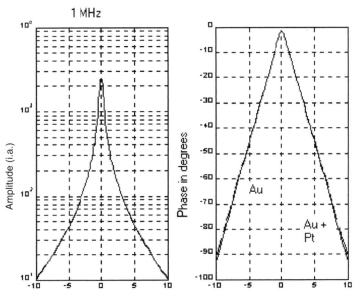

Fig. 4.7. Amplitude and phase versus the distance of the heat source of a layered sample (10 μm Au/Platinum) and a bulk gold sample at 1 MHz.

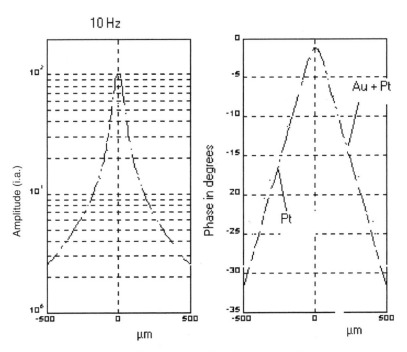

Fig. 4.8. Amplitude and phase versus the distance of the heat source at low frequency (10 Hz) of a layer sample (10 μm Au/platinum).

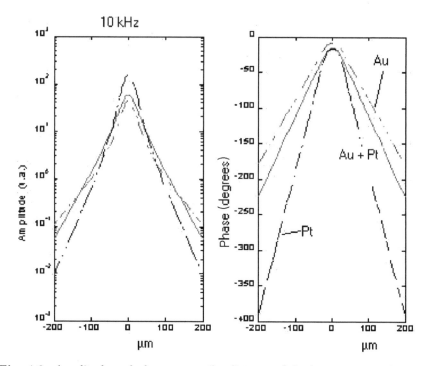

Fig. 4.9. Amplitude and phase versus the distance of the heat source at intermediate frequency (10 kHz) of a layer sample (10 μm Au/platinum), bulk gold sample and bulk Pt sample.

At high frequency, the thermal wave propagates only in the first layer and we can extract the thermal diffusivity of the first layer (Fig. 4.7). At very low frequency, the main contribution is the contribution of the bulk (Fig. 4.8), whereas the intermediate frequency experiments allow determining the effusivity ratio of the layer and the bulk (Fig. 4.9).

To do that, we have to fit the experimental data taken on the whole frequency range with the help of a best-fit program in which three parameters have to be determined: the thermal diffusivities of the layer, the substrate, and the ratio of the effusivities, for instance.

4.3 Photothermal Experiments on Complex Materials at Millimeter Scale

4.3.1 Determination of the Thermal Diffusivity with the Mirage Experiment

With respect to measuring the thermal diffusivity of a sample, we have to explore the heated area around the heat source created by the absorption of the intensity modulated excitation beam (50- to 100-μm-diameter Argon laser spot). In a mirage experiment [1], the measurement of this surface temperature variation is done through the measurement of the refractive index gradient associated with the temperature gradient developed above the heated area of the sample. The mirage cell [2] is a setup, which allows the measurement of very small deviation angles of a laser beam. In Fig. 4.10, we can see the compact (30-cm-long) block; a He-Ne laser is in its bottom. After two reflections, the laser beam is focused on the sample surrounding with a 7-cm focal length, corresponding to a 100-μm beam waist. The deviation angle is measured with a position sensor connected to a lock-in amplifier. This setup is easily photon-noise-limited, that is, the smallest deviation angle we can measure is on the order of 10^{-10} radian/Hz$^{1/2}$. This very small angle corresponds to 10^{-5} degrees/Hz$^{1/2}$ temperature variation on the sample surface. The modulation frequency range is on the order of a few kHz; above this frequency the thermal diffusion length in air is too small to compare to the diameter of the probe beam. To scan the heated area, we use computer-controlled step motors, which move the mirage cell at a constant distance $(100\,\mu m)$ above the sample surface. The extent of the heated area explored (a few millimeters) is dependent on the diffusivity of the sample and the modulation frequency. The phase of the mirage signal is a pertinent parameter to record, because it is less sensitive to the imperfections of the sample surface. But the amplitude has to be carefully examined in order to correlate the results obtained with the phase measurements. To determine the thermal diffusivity exactly, we have to solve the heat diffusion equations in all the media (air, coating, sample, etc.) and to calculate the deviation angle of the probe beam crossing the resulting temperature field in air. A best fit of our experimental data allows determination of the thermal diffusivity with a precision of about 10%.

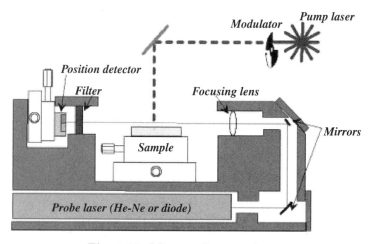

Fig. 4.10. Mirage cell setup scheme.

If we compare the conventional flash measurement with our mirage experiment, we can underline the following points:

- In the two experiments the sample surface has to be coated with an opaque layer for the photothermal excitation.
- The size (thickness and width) of the sample has to be perfectly determined in the case of the flash technique; in a mirage experiment well-defined size is not required.
- It is straightforward, by varying the modulation frequency, to measure the thermal diffusivity at various scales in the sample and then to get information from micrometer to millimeter scales.

Note finally that the two experiments are related by a Fourier transform, because the illumination is a light pulse in the flash laser technique and an intensity-modulated beam in the mirage technique.

4.3.2 Thermal Diffusivity Determination on CVD Diamond Samples

When the thickness of the CVD diamond films reaches a few hundred micrometers, the thermal quality of the sample is the same as the type II natural diamond in spite of its polycrystalline microstructure. Figure 4.11 shows measurements obtained with our mirage setup on a free-standing CVD diamond layer of 330-μm thickness, produced at the Philips research laboratories in Aachen, Germany, by a microwave-assisted method and covered with a 80-nm gold layer. An exponential regression of the amplitude and a linear regression of the phase far from the center give a value for the thermal diffusivity of $D = 10.5 \pm 1.0\,\text{cm}^2/\text{s}$. We have investigated a lot of polycrystalline diamond samples with thickness between one micrometer and a half millimeter, and we have clearly demonstrated the correlation between the nondiamond-phase contents estimated by Raman spectroscopy and the grain size with the thermal diffusivity estimation [3].

4.3.3 Aluminium Nitride Ceramics

Another interesting result is to know why the thermal conductivity of commercially available AlN ceramics ranges between 90 and 190 W/mK, whereas the thermal conductivity of a pure single crystal attains 320 W/mK. We have investigated and summarized the thermal diffusivity of various commercially available AlN substrates, the surface percentage of secondary phase deduced from image analysis, and atomic oxygen content in AlN grains as shown in Table 4.2. From the table, it seems difficult to clearly demonstrate the pertinent parameter we have to choose to get the best ceramic. Another important parameter we have to take into consideration is the morphology and thermal

Table 4.2. Comparison of thermal diffusivity and oxygen concentration in AlN grains and secondary phase percentage deduced from image analysis

Sample	A	B	C	D	E	F	G
D cm^2/s	0.35	0.4	0.5	0.57	0.68	0.78	0.95
Surface % of secondary phase	12	10	9	4	6	6	6
Oxygen atomic % in AlN grain	1%	0.95%	0.65%	0.65%	0.95%	0.9%	0.5%

properties of the secondary phase. To achieve this study, the local properties of both the AlN grains and the secondary phase have to be very well known.

The results at micrometer scales with our photothermal microscope are provided in the next section.

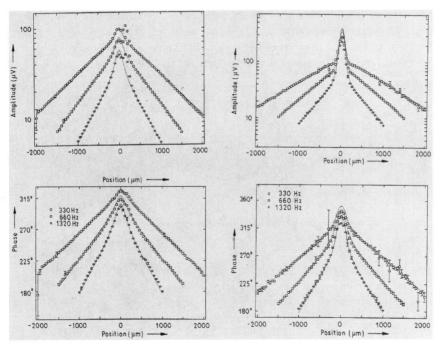

Fig. 4.11. Typical mirage experiment results: amplitude (top) and corresponding phase (bottom) obtained on Polycrystalline CVD 330-μm-thick diamond at different modulation frequencies (330 Hz, 660 Hz, and 1320 Hz). Left and right columns are the results obtained on the two sides (growth side and substrate side) of the polycrystalline CVD sample. The thermal diffusivity of this sample is evaluated to 10.5 ± 1.0 cm^2/s.

4.3.4 Silicon Nitride Ceramics

The thermal conductivities of Si_3N_4 ceramics are lower than those of AlN ceramics. But Si_3N_4 ceramics are very interesting for their excellent mechanical properties complementing their good thermal conductivity. Recently, a Si_3N_4 ceramic with highly anisotropic microstructure with large elongated grains uniaxially oriented along to the casting direction was developed by tape casting with seed particles. Mirage experiments [5] have clearly demonstrated that the sample is thermally anisotropic. The thermal diffusivity along the tape-casting direction $(0.54\,cm^2/s)$ is two times larger than the diffusivity along the two perpendicular directions $(0.35, 0.26\,cm^2/s)$. In order to understand the origin of this macroscopic thermal anisotropy, local determination of the thermal diffusivity is needed; this is discussed in the next section.

4.3.5 Thermal Heterogeneïty Evidence on Diamond Samples

A low-frequency mirage experiment can also give significant information on the microstructural heterogeneity of CVD polycrystalline samples. When comparing the curves measured on the substrate side (right) and on the growth side (left) of Fig. 4.11, we observe a much more pronounced maximum around the excitation on the substrate side than on the growth side. In Fig. 4.12,

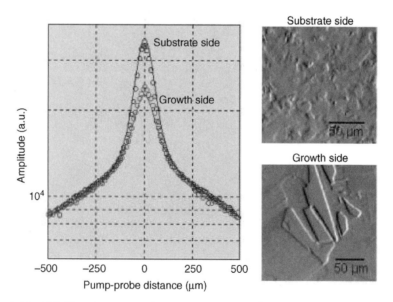

Fig. 4.12. 330-Hz mirage experiment on a 350-μm-thick sample. The central part of the amplitude given in arbitrary units of the experiment done on the substrate side is larger than when done on the growth side; this is a clear signature of the presence of a poor conductive layer on the substrate side.

we can see a superposition of the two amplitude measurements. This is the typical signature of an inhomogeneous sample; the high grain boundary density at the substrate side leads to a lower local diffusivity than at the growth side. The lines drawn in the plots represent a fit of the calculated mirage signal for a system of two layers with thickness and diffusivities of $l_1 = 30\,\mu\text{m}$, $D_1 = 1.0\,\text{cm}^2/\text{s}$ (substrate side) and $l_2 = 300\,\mu\text{m}$, $D_2 = 11.5\,\text{cm}^2/\text{s}$ (growth side). We note that the fit is predominantly sensitive to the thermal thickness of the layers. To get more precise inversion we have to do more experiments at various frequencies to probe the poor thermal conductor layer. Nevertheless, there can be no doubt about the presence of a poorly conducting layer near the substrate. Only high-frequency experiments can give more thermal information concerning this layer.

4.4 Photothermal Experiment at Microscopic Scale

4.4.1 Photothermal Microscope

Figure 4.13 shows the photoreflectance microscope built in our laboratory. The physical principle of this setup is the following: the optical reflection coefficient R is sensitive to the temperature, and a low-intensity DC probe beam flux impinging a sample surface whose temperature varies from T to $T + \Delta T$ is reflected with an optical reflection coefficient varying from R to $R + \Delta R$. The measurement of ΔR allows measurement of the temperature variation ΔT. The sample is heated by an intensity-modulated focused excitation spot. The temperature modulation is transferred via the refractive index and the reflectance to the reflected probe intensity. The amplitude and phase are then measured by a lock-in amplifier. To obtain spots as small as a few micrometers, we use a commercial microscope to focus the heating beam (Argon laser: 514 nm) as seen in Fig. 4.13. The probe beam is produced by a 670-nm laser diode. After reflection, the probe beam is received on a silicon photodiode coupled with a high-frequency lock-in amplifier. This experiment requires high modulation frequencies to prevent heat from diffusing over a large area if an experiment on a small area or if a photothermal image of good resolution is needed. It is possible to measure variations of ΔR as small as $10^{-6}\,R$ with a frequency range up to 20 MHz [6], [7], [8]. Note that our setup allows scanning the sample surface with the probe beam set on a fixed point of the surface while the excitation beam is scanned around it. This configuration allows measurements on poorly polished samples while the inverse configuration (heating spot fixed and scans with the probe beam) requires a good-optical-quality sample surface. In view of the reciprocity theorem for Green's functions of problems with homogeneous boundary conditions the detected signal would remain unchanged. Moreover, photothermal images of the sample surface with or without the two beams superimposed are obtained with a microcontrol actuator having a step resolution of $0.1\,\mu\text{m}$.

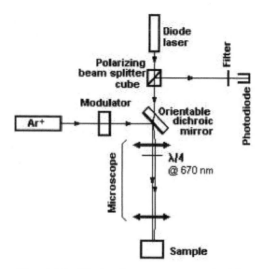

Fig. 4.13. Photoreflectance setup principle.

4.4.2 Thermal Diffusivity Measurement at a Single Grain Scale

4.4.2.1 AlN Ceramics. In the case of ceramic materials, we have already mentioned that the prediction of the macroscopic thermal behavior is related to the values of the local thermal determinations. In AlN ceramics with Y_2O_3 addition, the secondary phase often consists of yttrium-aluminates with low thermal conductivity. Figure 4.14 shows the photothermal signal phase maps

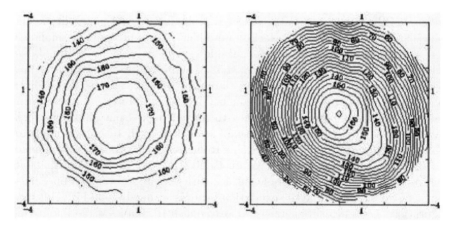

Fig. 4.14. Local high-resolution diffusivity measurement in an AlN-based Y_2O_3-sintered ceramic, coated with a 70-nm gold film. Phase contour lines in an AlN grain (left) and the Y_2O_3 intergranular phase (right), showing the much lower diffusivity of Y_2O_3. Modulation frequency: 1 MHz; length unit on both axes: 1 micrometer. The thermal diffusivity of the AlN grain is $0.4\,\mathrm{cm^2/s}$ and that of the secondary phase is ten times smaller.

of an AlN grain and the intergranular phase. The contours are very closely spaced on the intergranular phase, which corresponds clearly to a poor thermal conductor.

4.4.2.2 Si_3N_4 Ceramics.   Understanding the origin of the thermal anisotropy of the tape-casted Si_3N_4 ceramics is of great interest [5], [10]. Figure 4.15 shows the dependence of the amplitude and phase of the thermoreflectance signal on a Si_3N_4 grain (100 µm in length (*c*-axis direction) and 17 µm in width (*a*-axis direction)) as a function of the separation of the two spots. This figure shows strong thermal anisotropy inside the grain, depending on crystal axes. The principal thermal diffusivities were determined from the result of Fig. 4.15 and were 0.84 cm²/s for the *c*-axis of Si_3N_4 crystal and 0.32 cm²/s for

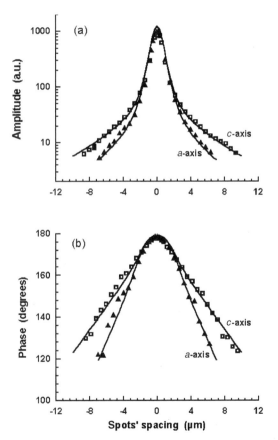

Fig. 4.15. Dependence of the amplitude (a) and of the phase (b) on the heating and probe spots' spacing, along the *a*-axis or *c*-axis, for the widest Si_3N_4 grain in our sample. Modulation frequency is 250 kHz. Experimental points are (triangles and squares and theoretical best fit are the solid lines). The thermal diffusivities in the two directions are 0.84 (*c*-axis) and 0.32 (*a*-axis) cm²/s.

the a-axis. The corresponding thermal conductivities were 180 and 69 W/mK. Measurements in grains of different sizes have shown the anisotropic diffusivities of the grains to be independent of the size of the grains. Thus, the thermal anisotropy is an intrinsic property of Si_3N_4 grains.

4.4.3 Photothermal Imaging

Photothermal microscopy is also capable of creating thermal images with micrometer resolution. Figure 4.16 is an example of such micrometer-scale resolution on carbon-carbon composite. We can clearly see how the heat diffuses when a heating point source is created in the center of the image. The composite exhibits a strong anisotropy in the heat diffusion due to its lamellar structure.

4.4.4 Thermal Barrier Evidence on AlN Ceramics

An additional capability of our setup is the detection of defects in the sample. If the defects or heterogeneous structures are parallel to the sample surface, 1D photothermal experiments versus the modulation frequency associated to an inversion procedure is a convenient method to detect them. It is also possible to detect the presence of vertical or slanted thermal barriers by scanning the sample with the pump spot and the probe spot superimposed in a 3D geometry. The thermal barriers prevent the heat from diffusing with a cylindrical geometry and are then associated with an increase in the amplitude and a variation in the phase. Figures 4.17, 4.18, and 4.19 are examples of thermal

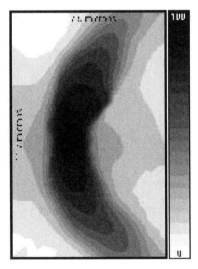

Fig. 4.16. Thermal anisotropy evidence of a carbon-carbon composite at the micron scale. Modulation frequency: 1 MHz—phase map around a point source.

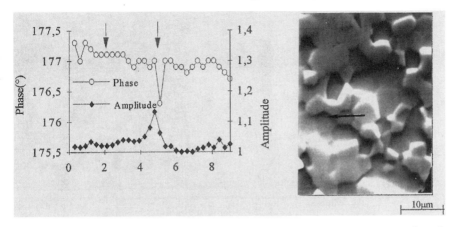

Fig. 4.17. 1-MHz modulation frequency photoreflectance signal associated with the scanning of two intergranular phases along the line on the SEM image. The scan starts in one grain of AlN, crosses the secondary phase, and finishes in another grain. Because the secondary phase is a poor thermal conductor the amplitude of the photothermal signal is higher.

barrier between two AlN grains, revealed by a photoreflectance experiment run at 1 MHz with the two beams superimposed [4], [9]. The white parts in the SEM images of the samples correspond to the secondary phase, and the gray area corresponds to AlN grains.

Modeling this kind of experiment requires the resolution of the heat diffusion equation without any cylindrical geometry. When the barrier is vertical, it is possible to solve the equation analytically [11], but when the barrier is slanted we have to use a numerical approach [12]. It is very difficult in this kind of experiment to estimate quantitatively the behavior of the thermal barrier, because it is difficult to guess the structure beneath the surface.

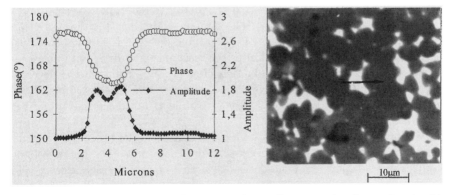

Fig. 4.18. 1-MHz modulation frequency photoreflectance signal associated with the scanning of a thick intergranular phase along the line on the SEM image.

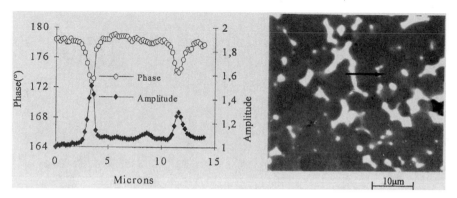

Fig. 4.19. 1-MHz photoreflectance signal associated with the scanning of three grains along the line on the SEM image. Note that grain boundary on the left does not exhibit any thermal signature.

Thus, in this study, we have tried to take into account macroscopic thermal diffusivity, percentage and distribution of the secondary phase, local diffusivity in each phase, and presence of thermal barriers at the grain. We have shown [9] that it is possible to predict the effective thermal conductivity if a random set model describes the distribution of the phases, if the local conductivity is well known and if the thermal continuity at the grain boundaries is effective. If these three conditions are fulfilled for one sample under investigation, we have effectively calculated two bounds between which the experimental value has been found [9]; another AlN sample cannot even be modeled. In fact, this sample shows many thermal barriers at grain boundaries, and a three-phase model including a third phase at the grain boundaries would be necessary to complete the simulation.

4.4.5 Very Thin Layer Thermal Property Determination

Another feature of our experiments is the ability to determine thermal properties of very thin layers [13], [14]. The samples under investigation were thin layers (100 to 400 nm thick) YBaCuO deposited on various substrates (ZrO_2, $LaALO_3$, and $SrTiO_3$). Thermal interface resistance and diffusivity measurements on thin high Tc superconducting films are important for the development of thin film devices such as photodetectors. The response time of the bolometer is actually dominated by the rate of heat transfer in the supporting substrates.

We have conducted photothermal experiments on a large scale of frequencies with our photothermal microscope. Note that the optical quality of the samples was adequate and the optical reflection coefficient was large enough to complete the experiments without any coating. Moreover, the optical reflection coefficient is around 10%, but the dR/dT is around 10^{-2}, which allows us to easily record the data.

4.4.5.1 Thermal Diffusivity Determination of the Substrate. In order to have
a precise determination of the thermal properties of the substrates, we first
did measurements on the substrates coated with a thin (80–90 nm) gold layer.
Figure 4.20 shows the experimental data and the best fit we have done on
four experiments at different frequencies. In this fit we have simultaneously
estimated the Au film diffusivity, the substrate diffusivity, and the thermal
boundary resistance. We have to take into account the thermal barrier, which
is present at the interface to fit our data. A multiparameter fitting procedure
has minimized the square variance between the measured and the calculated
phases. The experimental parameters, such as the diameters of the two (excitation and detection) lasers, were carefully measured and considered as fixed
parameters in the fitting procedure. Let us underline that at low frequency

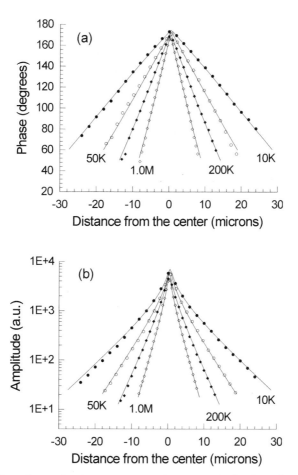

Fig. 4.20. Photothermal microscope experiments and the best-fit curves (solid
lines) on 89-nm Au deposited on Zirconia substrate at 4 modulation frequencies
(10 kHz, 50 kHz, 200 kHz, and 1 MHz).

Table 4.3. Experimental determination of the thermal properties of the YBCO samples deposited on three substrates [13], [14]

Film	Substrate	D_{th} Film cm^2/s	D_{th} Substrate cm^2/s	Rth (m^2K/W)
Au (89 nm)	ZrO_2	0.96	0.010	$0.4\ 10^{-8}$
Au (78 nm)	$LaAlO_3$	1.0	0.040–0.044	0.8–$1.0\ 10^{-8}$
YBCO (400 nm)	ZrO_2	0.019	0.010	$2.5\ 10^{-7}$
YBCO (130 nm)	ZrO_2	0.031	0.010	$2.2\ 10^{-7}$
YBCO (300 nm)	$LaAlO_3$	0.032	0.042	$2.0\ 10^{-7}$
YBCO (300 nm)	$SrTiO_3$	0.030–0.033	0.035	2.5–$1.6\ 10^{-7}$

the substrate diffusivity is dominating, while at high frequency the influence of the film diffusivity and the thermal resistance increase. It is then very important to conduct experiments at different frequencies. The results are given in Table 4.3.

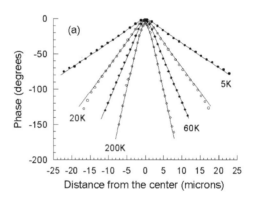

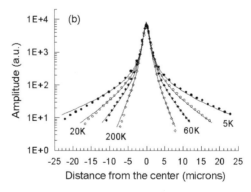

Fig. 4.21. Photothermal microscope experiments and the best-fit curves (solid lines) on 300-nm YBaCuO deposited on $LaAlO_3$ substrate at 4 modulation frequencies (5 kHz, 20 kHz, 60 kHz, and 200 kHz).

4.4.5.2 Determination of the YBaCuO Layer Thermal Diffusivity and of the Thermal Interface Resistance.   Once the method is validated on gold layers, we proceed to the investigation of the YBaCuO samples. The experimental data were taken at at least three frequencies (usually four) and then the same fitting procedure was done. Figure 4.21 shows the experimental data and the results of the fitting procedure on YBaCuO layer deposited on $LaAlO_3$.

Table 4.3 gathers all the results on all the samples we investigated. The thermal resistance determination can be more precise if the layer and substrate properties are comparable. But it is very important to determine simultaneously the three parameters.

4.5 Conclusion

The setups developed in our laboratory are unique tools to measure thermal properties at different spatial scales without much metallographic preparation. The mirage technique is possible to investigate the thermal behavior at millimeter scale, whereas photothermal microscopy is possible to investigate at microscopic scale. Combination of the information obtained at these scales and statistical morphological data obtained by image analysis makes it possible to analyze the macroscopic thermal behavior of bulk material and films.

Acknowledgments

The author gratefully acknowledges the efficient work of many colleagues and students who are at the origin of the results presented in this chapter: Karsten Plamann, who did all the experiments on the CVD diamond samples; Catherine Pelissonnier-Grosjean, Alain Thorel, Dominique Jeulin, and Lionnel Pottier who have achieved the interesting work on the AlN ceramics; Bincheng Li and Jean Paul Roger for their contribution on Si_3N_4 ceramics and YBaCuO films; and Stephane Hirschi and Julien Jumel for the thermal image of carbon-carbon composites.

References

[1] A. C. Boccara, D. Fournier, and J. Badoz, *Appl. Phys. Lett.* **36**, 130 (1979).

[2] F. Charbonnier, and D. Fournier, *Rev. Sci. Instrum.* **57**, 1126 (1986).

[3] K. Plamann and D. Fournier, *Phys. Stat. Sol.* **154**, 351 (1996).

[4] C. Pelissonnier-Grosjean, L. Pottier, D. Fournier, and A. Thorel, in *Proc. Fourth Euro Ceramics* **3**, 413 (1995).

[5] B. Li, L. Pottier , J. P. Roger, D. Fournier, K. Watari, and K. Hirao, *J. Eur. Ceram. Soc.* **19**, 1631 (1999).

[6] A. Rosencwaig, J. Opsal, W. L. Smith, and D. L. Willenborg, Appl. *Phys. Lett.* **46**, 1013 (1985).

[7] L. J. Inglehart, A. Broniatowski, D. Fournier, A. C. Boccara, and F. Lepoutre, *Appl. Phys. Lett.* **56**, 18 (1990).

[8] L. Pottier, *Appl. Phys. Lett.* **64**, 1618 (1994).

[9] C. Pelissonnier-Grosjean, D. Jeulin, L. Pottier, D. Fournier, and A. Thorel, in *Proc. Euro Ceramics V Engineering Materials* **132–6**, 623 (1997).

[10] K. Hirao, K. Watari, M. Brito, M. Toriyama, and S. Kanzaki, *J. Amer. Ceram. Soc.* **79**, 2485 (1996).

[11] F. Lepoutre, D. Balageas, P. Forge, S. Hirschi, J. L. Joulaud, D. Rochais, and F. C. Chen, *J. Appl. Phys.* **78**, 2208 (1995).

[12] A. M. Mansanares, T. Velinov, Z. Bozoki, D. Fournier, and A. C. Boccara, *J. Appl. Phys.* **75**, 1 (1994).

[13] B. Li, L. Pottier, J. P. Roger, and D. Fournier, *Thin Solid Films* **352**, 91–6 (1999).

[14] B. Li, J. P. Roger, L. Pottier, and D. Fournier, *J. Appl. Phys.* **86**, 5314–16 (1999).

5

Fabrication of High-Thermal-Conductivity Polycrystalline Aluminum Nitride: Thermodynamic and Kinetic Aspects of Oxygen Removal

Anil V. Virkar and Raymond A. Cutler

It is now well recognized that the principal factor that lowers the thermal conductivity of polycrystalline AlN well below its theoretical value of ~320 W/mK at room temperature is the presence of dissolved lattice oxygen. Covalently bonded AlN is difficult to fabricate by pressureless sintering due to very low diffusion coefficients. It is also now well-established that certain additives, predominantly oxides, facilitate densification of AlN by liquid phase sintering, and in the process lead to the purification of AlN lattice—a prerequisite to attaining high thermal conductivity. The additive must have the ability to form a liquid phase, which facilitates sintering at the processing temperature. The additive must also have a strong enough affinity for aluminum oxide (Al_2O_3), which is the form in which oxygen is dissolved in AlN, so that various aluminates can be formed as secondary phases—effectively scavenging Al_2O_3 and purifying the lattice. The thermodynamic considerations relate to the affinity between Al_2O_3 and the additive to form aluminates, which can be described in terms of the standard free energy of formation, ΔG^o, of the respective aluminate. The greater the $|\Delta G^o|$, with $\Delta G^o < 0$, the greater is the ability of the additive to scavenge Al_2O_3. The kinetics relate to various processes, such as the sintering kinetics and the kinetics of the removal of dissolved oxygen from within the grains to grain boundaries. In addition, other kinetic processes involve changes at the microstructural level, such as changing the wetting characteristics of secondary grain boundary phases and the occurrence of grain growth. This chapter presents a brief review of the role of dissolved oxygen on the thermal conductivity of AlN and a brief review of the role of sintering and processing procedures used in the fabrication of high-thermal-conductivity AlN ceramics. The focus will be on using the lattice model of thermal conductivity of AlN, the role of thermodynamics and phase equilibria in the purification of AlN lattice, and the kinetics of various processes central to the fabrication of high-thermal-conductivity AlN ceramics.

5.1 Theoretical Basis

Aluminum nitride (AlN) in a polycrystalline form has been investigated for application as a substrate material for microelectronics applications due to its attractive properties, including (1) a moderately low dielectric constant and a low loss tangent, (2) a coefficient of thermal expansion similar to that of silicon, (3) high electrical resistivity (electrical insulator), and (4) high thermal conductivity when appropriately processed. AlN is covalently bonded and has the wurtzite structure. Slack and coworkers identified AlN as a potential material for application in the microelectronics industry for the aforementioned reasons [1], [2], [3]. All other properties, such as the coefficient of thermal expansion, elastic constants, and heat capacity, are not strong functions of the processing history as long as the final material is of high density (low porosity). Such, however, is not the case with thermal conductivity. It is now well known that depending on how the material is processed, the thermal conductivity, κ, of AlN can vary over a very wide range, from as low as $40\,\mathrm{W/mK}$ to more than $270\,\mathrm{W/mK}$ at room temperature. It is this great variability in thermal conductivity as a function of processing parameters and the potential for application in the microelectronics industry that has attracted the attention of researchers.

Slack and coworkers were the first to recognize that the principal factor that determines the thermal conductivity of AlN at ambient temperature is related to impurities dissolved in the lattice [1], [2], [3]. This is because the main mode of heat conduction in AlN, which is an electrical insulator, is by lattice vibrations—phonons. At a given temperature, the highest thermal conductivity that an electrically insulating material can exhibit is governed by the phonon mean free path, which is dictated by phonon-phonon interactions. Impurities, however, can scatter phonons and effectively lower the thermal conductivity of AlN. In the simplest description of thermal conduction processes wherein the principal mode of heat transmission is by phonons, the thermal conductivity can be adequately described by an equation of the form [4]

$$\kappa = \frac{1}{3}Cv\ell, \qquad (5.1)$$

where C is the heat capacity, v is the lattice velocity, and ℓ is the phonon mean free path. Neither the heat capacity, C, nor the lattice velocity, v, is sensitive to the possible presence of impurities. However, impurities can significantly alter the phonon mean free path, ℓ, and thus effectively influence κ. Slack and coworkers further proposed that it is the presence of oxygen dissolved in the AlN lattice that serves as an effective phonon scatterer [3]. More accurately, it is the vacant aluminum site that forms on the dissolution of aluminum oxide, Al_2O_3, in AlN lattice which is the culprit. AlN invariably contains some dissolved oxygen in its lattice due to the manner in which it is often made. Also the surface of fine powders readily oxidizes when exposed to air to form Al_2O_3. Some of the surface Al_2O_3 can dissolve into the AlN lattice at the high temperatures required for processing. The incorporation of Al_2O_3 in the AlN lattice can be described by the following defect reaction:

$$Al_2O_3 \longrightarrow 2Al_{Al} + V_{Al} + 3O_N, \qquad (5.1a)$$

where aluminum atoms occupy aluminum sites, oxygen substitutes for nitrogen, and V_{Al} denotes a vacant aluminum site. It is the mass difference between a regular Al atom and its vacancy, V_{Al}, that makes the vacancy a strong phonon scatterer. That is, oxygen has an indirect effect in reducing κ, and increased κ in an AlN ceramic can be realized if the lattice is purified by removing as much of the dissolved Al_2O_3 as possible. The ability of an impurity to scatter phonons can be described in terms of a parameter known as the phonon-scattering cross section, Γ, and is given by [5], [6]

$$\Gamma = x_s(1 - x_s)\left[\left(\frac{\Delta M}{M}\right)^2 + \chi\left(\frac{\Delta\delta}{\delta}\right)^2\right], \tag{5.2}$$

where M is the atomic weight of the host atom, ΔM is the difference in the atomic weights between the host and the impurity species occupying the site normally occupied by the host atom, δ is the atomic diameter of the host atom, $\Delta\delta$ is the difference between the atomic diameters of the host and the impurity, χ is a dimensionless constant, and x_s is the site fraction occupied by the impurity. For very small concentrations of the impurity, such as would be the case with dissolved oxygen, Eq. (5.2) may be approximated by

$$\Gamma \approx x_s\left[\left(\frac{\Delta M}{M}\right)^2 + \chi\left(\frac{\Delta\delta}{\delta}\right)^2\right]. \tag{5.3}$$

With Al_2O_3 dissolved in AlN, the $(\Delta M/M)^2 \approx 0.02$ for oxygen (atomic weight $= 16$) occupying a nitrogen (atomic weight $= 14$) site, which is very small. However, for an empty aluminum site, $(\Delta M/M)^2 = 1$, which is equivalent to a vacancy occupying a site normally occupied by aluminum. It is for this reason that oxygen as an impurity is effective in scattering phonons in an indirect manner. That is, had aluminum oxide existed as AlO instead of Al_2O_3, its effectiveness as a phonon scatterer would have been very small. Slack and coworkers have examined the effect of oxygen as an impurity on the thermal conductivity of AlN. They demonstrated that the thermal conductivity of polycrystalline AlN decreases with increasing dissolved oxygen concentration and obeys a relation of the form

$$\frac{1}{\kappa} \approx \frac{1}{\kappa^o_{AlN}} + \alpha\frac{\Delta n}{n_o}, \tag{5.4}$$

where κ^o_{AlN} is the thermal conductivity of pure AlN, n_o is the number of nitrogen atoms per unit volume, Δn is the number of oxygen atoms per unit volume, and α is a constant. At room temperature, κ^o_{AlN} is about 320 W/mK and the constant α was determined to be 0.43 mK/W. This suggests that at room temperature, one may use Eq. (5.4) to estimate the oxygen concentration by rearranging it as follows:

$$\frac{\Delta n}{n_o} \approx \frac{1}{\alpha}\left(\frac{1}{\kappa} - \frac{1}{\kappa^o}\right). \tag{5.5}$$

Equation (5.4) describes the dependence of thermal conductivity as a function of oxygen content at a given temperature. Abeles has examined the thermal conductivity of materials as a function of both the temperature and impurity-induced phonon scattering [5]. The general form of the equation is given by

$$\kappa \approx \frac{1}{A\sqrt{\Gamma T} + BT}, \tag{5.6}$$

where the constants A and B are given by

$$A = 9.67 \times 10^2 \gamma_1 \sqrt{(1 + 5\varepsilon/9)}\sqrt{M}\beta^{-2}\delta^3 \tag{5.7}$$

and

$$B = 7.08 \times 10^{-4}\gamma_1^2 (1 + 5\varepsilon/9)\sqrt{M}\beta^{-3}\delta^{7/2}, \tag{5.8}$$

where γ_1 is the aharmonicity factor, ε and β are constants, and δ^3 is the atomic volume. Equation (5.6) may be rearranged as

$$\frac{1}{\kappa} = A\sqrt{\Gamma T} + BT \tag{5.9}$$

or

$$\frac{1}{\kappa\sqrt{T}} = A\sqrt{\Gamma} + B\sqrt{T}. \tag{5.10}$$

The significance of Eq. (5.10) from the standpoint of application to experimentally measured thermal conductivity is that a plot of $1/\kappa\sqrt{T}$ vs. $\sqrt{T}$ should be a straight line, with $A\sqrt{\Gamma}$ as the intercept and B as the slope. Equation (5.6) is thus useful in analyzing experimentally measured thermal conductivity as a function of temperature. Equation (5.6) suggests that at very low temperatures (but still above the Debye temperature), the term in the denominator under the square root is dominant, and the thermal conductivity is dictated by phonon-impurity scattering. At high temperatures, the term in the denominator that is linear in temperature is dominant, and the thermal conductivity is dictated by phonon-phonon scattering. Equation (5.6) thus is useful for the analysis of data to identify the role of both dissolved oxygen and temperature.

5.2 Procedures for the Fabrication of High-Thermal-Conductivity Aluminum Nitride Ceramics

As stated earlier, because AlN is a very refractory material with covalent bonding, diffusion coefficients are rather low even at elevated temperatures. This makes pressureless densification by a solid-state method difficult, although this has been recently achieved using carefully prepared fine AlN powders. The most direct method of sintering AlN is by adding other materials as sintering aids to form a liquid phase at the sintering temperature. Many rare earth oxides and alkaline earth oxides are known to form eutectic liquids with Al_2O_3. This fact has been long recognized and forms the basis of sintering processes for the fabrication of high-density AlN ceramics [7], [8], [9], [10], [11]. The pioneering work of Komeya and coworkers has demonstrated that AlN can be sintered to a near theoretical density by adding various additives, typically oxides [7], [8]. The oxide additives serve two important functions: (1) They react with surface Al_2O_3 to instantaneously form a liquid phase at elevated temperatures facilitating rapid densification, and (2) they effectively serve to remove dissolved oxygen from the lattice and thereby enhance κ. Thus, the addition of rare earth and alkaline earth oxides assists in densification, and in the process purifies the AlN lattice such that the resulting multiphase mixture (purified AlN and aluminates) actually has a considerably higher thermal conductivity than the original AlN, despite the fact that aluminates are generally poor thermal conductors. The reason is that oxygen dissolved in AlN is far more detrimental to the thermal conduction process, because it lowers the phonon mean free path (increases the phonon scattering cross section). By contrast, oxygen present in secondary phases, such as aluminates, has a relatively weak effect on thermal conductivity as long as the AlN phase is contiguous. The thermal conductivity of a two-phase mixture can usually be described in terms of an appropriate rule of mixtures. Not all additives are expected to be equally effective in purifying AlN. For example, the phase diagram between Al_2O_3 and the oxide additive determines the effectiveness of an additive as a sintering aid, depending on the temperature at which a liquid phase forms. Also, the effectiveness of an additive in oxygen scavenging depends on the thermodynamic affinity of the oxide additive for Al_2O_3. Finally, factors such as the kinetics of the overall process and the resulting morphology of grain boundary oxide phases determine the final properties of sintered AlN. In the remainder of this chapter, the effects of some thermodynamic, kinetic, and microstructural factors on the fabrication of high-thermal-conductivity AlN ceramics is discussed, and the thermal conductivity is examined in light of the lattice (phonon) model.

The typical procedure for the fabrication of polycrystalline AlN ceramics consists of the following steps:

1. Mix fine AlN powder with the powder of an additive, which may be a rare earth oxide or an alkaline earth oxide. The required amount of the additive depends on the purity of the AlN powder, especially its oxygen content.
2. Consolidate the powder in the form of discs, plates, etc. using die-pressing, tape-casting, or other green-forming methods.
3. After removing organics at a lower temperature, sinter and anneal in a reducing atmosphere. The typical sintering temperature is on the order

of 1850°C, and the typical sinter/annealing time is on the order of a few minutes to a few hours.

Various other sintering/annealing procedures are used depending on powder characteristics, the type and amount of the additive, and the desired final properties.

5.3 Phase Equilibria, Sintering, and Thermodynamic Considerations

Sintering is typically achieved by liquid phase sintering, wherein a liquid phase is formed between Al_2O_3 inherently present on the surface of a typical AlN powder and the additive. Over the range of typical sintering temperatures (~1650 to 1900°C), liquid phases form in alkaline earth oxide—Al_2O_3 systems and a number of rare earth oxide—Al_2O_3 systems. Also, several aluminates, usually exhibiting very narrow ranges of stoichiometry (line compounds), exist in these systems, the significance of which is discussed shortly. To date, high-thermal-conductivity AlN ceramics have been made using a number of oxide additives, such as Y_2O_3, Yb_2O_3, Dy_2O_3, Er_2O_3, Gd_2O_3, Nd_2O_3, and Sm_2O_3, as well as with CaO; and many other inorganic materials including some carbides and fluorides. The following discussion will be confined to oxides, in particular to Y_2O_3 for the purposes of illustration, although the general conclusions are applicable to essentially all other rare earth oxide–based and other oxide additive–based systems.

The system Al–Y–O–N can be represented as a quaternary isothermal section with end members AlN (or Al_4N_4)–YN (or Y_4N_4)–Y_2O_3 (or Y_4O_6)–Al_2O_3 (or Al_4O_6), as shown in Fig. 5.1. Also shown are the possible condensed two-phase and three-phase fields. All phases are assumed to have very limited stoichiometry ranges. Thus, each single-phase field is represented by a point, for example, the Al_4N_4 corner. Figure 5.1(a) shows the various phase fields at a temperature at which no liquid phase is present. Figure 5.1(b) shows the various phase fields at a temperature at which a liquid phase is present. The latter is the relevant phase diagram at the sintering temperature and may also represent conditions during subsequent annealing. In the figure, **Y** represents Y_2O_3 and **A** represents Al_2O_3. Thus, **YA** represents $Y_2O_3Al_2O_3$ or $YAlO_3$; the perovskite phase, $\mathbf{Y_3A_5}$, represents $(Y_2O_3)_3(Al_2O_3)_5$ or $Y_3Al_5O_{12}$, the yttrium aluminate garnet phase (YAG); and $\mathbf{Y_2A}$ represents $(Y_2O_3)_2Al_2O_3$ or $Y_4Al_2O_9$. The pseudobinary AlN–Al_2O_3 is represented by the left-hand vertical side of the quaternary in Figs. 5.1(a) and (b). The phase ALON has the approximate stoichiometry Al_3O_3N, which actually has a relatively wide stoichiometry range and should strictly be shown as a line along the Al_4N_4–Al_4O_6 join and as an area extending into the quaternary. The starting AlN powder thus can be represented (insofar as the overall composition is concerned) by a point on this line, close to the AlN (or Al_4N_4) corner. The

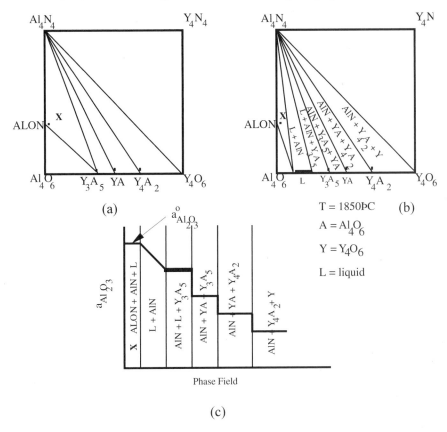

Fig. 5.1. (a) Schematic of the AlN–Al$_2$O$_3$–Y$_2$O$_3$–YN quaternary at a temperature at which no liquid phase is present; (b) schematic of the AlN–Al$_2$O$_3$–Y$_2$O$_3$–YN quaternary at a temperature at which a liquid phase is present; (c) schematic equilibrium activity of Al$_2$O$_3$, $a_{Al_2O_3}$, as a function position in a given phase field corresponding to Fig. 5.1(b).

addition of Y$_2$O$_3$ (or Y$_4$O$_6$) moves this point in the interior of the quaternary, close to the AlN corner, as the overall concentration of AlN is much greater than that of any of the other species. However, at equilibrium, the composition may lie in different triangles (three (condensed) phase fields) or in two-phase (solid) fields (characterized by lines separating three phase fields) or in a two-phase field with a liquid phase present, characterized by a near triangle. The Al-Y-O-N quaternary diagram is a four-component (C) system. In a phase field containing three condensed phases, the actual number of phases (P) is four (three condensed + gas). According to the Gibbs phase rule, the number of degrees of freedom that a system has is given by

$$F = C - P + 2. \tag{5.11}$$

If the temperature is fixed and the total pressure is fixed, then according to the Gibbs phase rule, a four-component system with four phases present (three condensed + gas) has no degrees of freedom (F) left. (The total pressure can be fixed by fixing, for example, nitrogen pressure, which fixes partial pressures of all other species.) In such a case, the chemical potentials or the thermodynamic activities of all species are fixed. If, on the other hand, there are two condensed phases, then $P = 3$ and the system has one degree of freedom, that is, $F = 1$. In such a case, the activities of species can vary across the two-phase field. (From here on, we will only state the number of condensed phases present. Thus, when we refer to a phase field as a two-phase field, the existence of the gas phase is implicitly assumed, and the system actually has three phases, that is, $P = 3$.) The implication is that in a three-phase field, the activity of Al_2O_3, $a_{Al_2O_3}$, is constant, regardless of where the composition is within a given three-phase field. The compositions of interest are very close to the AlN apex of the triangle. The activity of Al_2O_3 in AlN is also the same, because solid AlN is one of the phases corresponding to a given three-phase field. By contrast, the $a_{Al_2O_3}$ is not fixed in a two-phase field, characterized by a (near) triangle when a liquid is present (Fig. 5.1(b)) or essentially by a line when the two phases present are solid. The further implication is that the activity of Al_2O_3, $a_{Al_2O_3}$, changes abruptly across such a two-phase field. Figure 5.1(c) shows the dependence of $a_{Al_2O_3}$ as a function of composition, that is, as a function of the phase field the composition is in, for the quaternary shown in Fig. 5.1(b). Note the abrupt changes in $a_{Al_2O_3}$ as the composition shifts from one three-phase field to another three-phase field, across the dividing two-phase field indicated by a line. For the two-phase field comprising L + AlN, however, the $a_{Al_2O_3}$ varies smoothly, consistent with the Gibbs phase rule and the preceding discussion.

Referring to Fig. 5.1(b), note that as the amount of Y_2O_3 added is increased relative to Al_2O_3 inherently present in the starting AlN powder, the composition progressively shifts into three-phase fields containing AlN and aluminates richer in yttrium. With an increase in the Y_2O_3 content, the relative amount of low-thermal-conductivity secondary phases increases and the activity of Al_2O_3 in AlN decreases. This means, effectively, that the AlN lattice is increasingly purified. This has two effects: (1) an increase in the amount of secondary (aluminate) phases should lead to a decrease in thermal conductivity of the multiphase mixture; (2) purification of the AlN lattice should lead to an increase in the thermal conductivity. At small concentrations of Y_2O_3, as the amount of Y_2O_3 is increased, the increased purification of the AlN lattice leading to an increase in κ outweighs a decrease in κ due to the increase in the volume fraction of the aluminate phase. Thus, κ initially increases with increasing Y_2O_3 content. At large concentrations of Y_2O_3, a compromise between these two factors can lead to a maximum in thermal conductivity. This aspect is discussed later.

5.3.1 Free Energies of Formation and the Activity of Al_2O_3

The ability of a given oxide additive to scavenge Al_2O_3 out of AlN depends on the free energies of formation of the respective aluminates. In the Y_2O_3–Al_2O_3 system, for example, the formation of $YAlO_3$ can be described by the following reaction

$$\tfrac{1}{2}\,Y_2O_3 + \tfrac{1}{2}\,Al_2O_3 \longrightarrow YAlO_3 \qquad (5.11')$$

for which the standard free energy change is given by ΔG_{ii}^o. Because this reaction is known to occur over a wide temperature range, it is clear that $\Delta G_{ii}^o < 0$. If neither Y_2O_3 nor Al_2O_3 is present as a secondary phase and the only oxide phase present is $YAlO_3$, then at equilibrium we know that

$$a_{Al_2O_3} \times a_{Y_2O_3} = \exp\left[\frac{2\Delta G_{ii}^o}{RT}\right], \qquad (5.12)$$

where $a_{Y_2O_3}$ is the activity of Y_2O_3 and R is the universal gas constant. Similar equations can be written for other aluminates. Unfortunately, free energies of formation of the majority of the aluminates are generally not available, which precludes a quantitative determination of the activity of Al_2O_3. Nevertheless, thermodynamics provides a powerful scientific basis on which processing decisions for the fabrication of high-thermal-conductivity AlN can be made. For example, a schematic free energy–versus–composition diagram can be developed for a given system based on the knowledge of the phase diagram. Consider again the Y_2O_3–Al_2O_3 system, for which the phase diagram is known. In what follows, we restrict discussion to experimental conditions corresponding to all condensed phases being solid, for example, corresponding to Fig. 5.1(a), for the purposes of illustration. Thus, in this case, the changes in $a_{Al_2O_3}$ are expected to be abrupt across single-phase fields in the Y_2O_3–Al_2O_3 system (or across two-phase fields in the Y_2O_3–Al_2O_3–AlN system). Based on the fundamentals of phase equilibria and thermodynamic considerations, a schematic of the free energy–versus–composition diagram can be established. This is given in Fig. 5.2(a), in which $Y_2O_3(\mathbf{Y})$ is on the left side and $Al_2O_3(\mathbf{A})$ is on the right side. The vertical lines extending down from the horizontal line (axis) in Fig. 5.2(a) are measures of free energies of formation of the various aluminates in the Y_2O_3–Al_2O_3 system, namely, $\mathbf{Y_2A}$ (the same as $\mathbf{Y_4A_2}$), $\mathbf{YA}$, and $\mathbf{Y_3A_5}$. The straight lines joining them at the ends of these lines and forming an approximately concave curve upward is the free energy–versus–composition diagram at the temperature of interest (here corresponding to Fig. 5.1(a), for example). A tangent drawn to this curve at any point intersects the two vertical axes (corresponding to pure $\mathbf{Y}$ and pure $\mathbf{A}$). The intercept on the $\mathbf{A}$-axis is directly proportional to the logarithm of the corresponding activity of Al_2O_3, that is, $\ln a_{Al_2O_3}$. The free energy–versus–composition trace is not a smooth curve, because the phase diagram is characterized by the presence of line compounds and two-phase fields between them, rather it is made of segmented straight-line portions. (If the phase diagram were

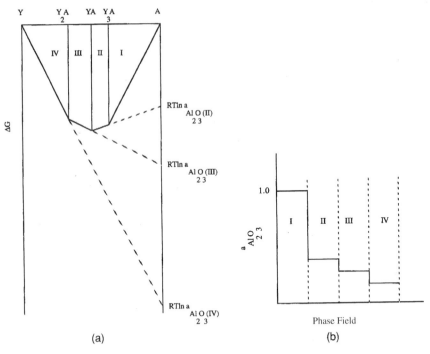

(a)   (b)

Fig. 5.2. (a) Schematic of the free energy of formation versus composition diagram for the Y_2O_3–Al_2O_3 system corresponding to Fig. 5.1(a); (b) schematic equilibrium activity of Al_2O_3, $a_{Al_2O_3}$, in the various three-phase fields corresponding to Fig. 5.1(a).

to correspond to a solid solution, then the free energy vs. composition trace would have been a smooth, concave-up curve.) Region III in Fig. 5.2(a), for example, is a two-phase field in the **Y-A** system (which, with the presence of AlN, would be a three-phase field) with the two phases being $\mathbf{Y_2A}$ and **YA**. An extension of the line joining the ends of the vertical lines corresponding to the free energies of formation of **YA** and $\mathbf{Y_2A}$ (effectively the tangent) to the **A**-axis intersects it at some place. This vertical intercept (which is negative) is given by $RT \ln a_{Al_2O_3(III)}$, from which the activity of Al_2O_3 can be estimated. In the presence of AlN, if the oxide phase equilibrium corresponds to region III, it means that the activity of Al_2O_3 dissolved in AlN is also given by $a_{Al_2O_3(III)}$. Returning to Fig. 5.2(a), note that for region I corresponding to the presence of $\mathbf{Y_3A_5}$ and **A**, the activity of Al_2O_3 is essentially unity—the same as for pure Al_2O_3. (Note, however, that the highest activity of Al_2O_3 in equilibrium with AlN is that corresponding to the AlN–ALON–$\mathbf{Y_3A_5}$ three-phase field (marked by X) in Fig. 5.1(a), which is less than unity. In Fig. 5.1(c), this activity is identified by $a^o_{Al_2O_3}$, where $a^o_{Al_2O_3} < 1$.) However, the activity of Al_2O_3, $a_{Al_2O_3}$, is much lower in region II, which represents the coexistence of $\mathbf{Y_3A_5}$ and **YA**. The $a_{Al_2O_3}$ continues to decrease as the **Y** content is increased, and the phase field progressively shifts from I to II to III and eventually to IV. Figure 5.2(b) shows the variation of the activity of Al_2O_3, $a_{Al_2O_3}$, as a function of the Y_2O_3 content, which is similar to Fig. 5.1(c). Clearly, when AlN is also present, the equilibrium $a_{Al_2O_3}$ corresponding to Al_2O_3 (or oxygen) dissolved in AlN follows the same trend. This shows that with the addition of **Y**, we expect the purification of AlN lattice to occur—a prerequisite to achieving high thermal conductivity.

The implication of the preceding discussion concerning the thermal conductivity of AlN, when Y_2O_3 is added as a scavenger, is that there must also be an abrupt change in thermal conductivity as the composition shifts from, say, region II $(AlN + \mathbf{Y_3A_5} + \mathbf{YA})$ to region III $(AlN + \mathbf{YA} + \mathbf{Y_2A})$. Experimental work by Jackson et al., who made AlN samples by adding various yttrium aluminates directly to AlN and then fabricating the samples, is consistent with this expectation [10]. Figure 5.3 shows the measured thermal conductivity of as-fabricated AlN samples made with no yttrium-containing

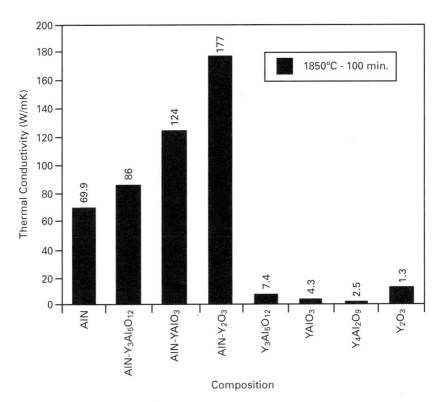

Fig. 5.3. Thermal conductivity comparison of hot-pressed AlN and AlN sintered with Y_2O_3 or yttrium aluminate additives and the thermal conductivity of individual aluminates.

phase added, with Y_3A_5 added, with YA added, and with Y added. Note that the thermal conductivity increases from ~70 W/mK for AlN with no Y added to ~86 W/mK with Y_3A_5 added to ~124 W/mK with YA added and finally to ~177 W/mK with Y added.

These results thus demonstrate the role of thermodynamics in achieving a high-thermal-conductivity AlN ceramic. A thorough quantitative analysis will have to be deferred until relevant thermodynamic data become available. In general, however, we do not expect great variability between the various rare earth oxides because the general features of their phase diagrams are similar, indicating that possible differences in free energies (among the various rare earth aluminates) are probably small. As a general rule, the magnitudes of free energies of formation of the various aluminates, ΔG^os, namely $|\Delta G^o|$, are expected to be modest (typically less than 100 kJ/mol). This is because when Al_2O_3 (an oxide) reacts with a rare earth oxide (RE_2O_3), no further valence changes are expected, as all cations are already in their fully oxidized states. Thus, the magnitude of the enthalpy of the reaction, which is primarily related to short-range interactions, is usually small, and so is the magnitude of the free energy of formation. Jackson et al. noted that the thermal conductivity of the final sintered AlN ceramic was less sensitive to which rare earth oxide was used [12]. (With the exception of CeO_2 and Eu_2O_3, using which dense samples could not be obtained [12].) Rather, the thermal conductivity was more sensitive to the amount of the rare earth oxide additive, consistent with the discussion on the role of thermodynamics and the processing history. We emphasize that even though the $|\Delta G^o|$ for the majority of the aluminates is expected to be small, the relative difference between two oxides can be significant. It can be shown that the equilibrium activity of $a_{Al_2O_3}$ is given by

$$a_{Al_2O_3} \approx \exp\left[\frac{\xi \Delta G^o}{RT}\right], \tag{5.13}$$

where ξ is a positive number whose magnitude depends on the stoichiometry of the aluminate formation reaction. We will assume ξ to be one in what follows. Then, for a ΔG^o of $-10\,\text{kJ/mol}$ and a processing temperature of 1800°C, the estimated $a_{Al_2O_3}$ is ~0.56. If, on the other hand, the ΔG^o is $-50\,\text{kJ/mol}$, the corresponding $a_{Al_2O_3}$ is ~0.055, that is, a purification by a factor of about 10. Indeed, it has been shown that with CaO as the additive, thermal conductivity of ~137 W/mK can be achieved. However, with MgO as the additive, the thermal conductivity achieved was only ~63 W/mK [13]. This is consistent with the measured free energies of formation of $CaAl_2O_4$ ($\sim -54\,\text{kJ/mol}$) and $MgAl_2O_4$ ($\sim -37\,\text{kJ/mol}$).

5.3.2 Thermodynamics of Oxygen Removal and the Analysis of Thermal Conductivity

Equation (5.6) describes the combined effect of phonon-phonon interactions and phonon-impurity interactions on the thermal conductivity.

Equation (5.10) suggests that a plot of $1/\kappa\sqrt{T}$ versus $\sqrt{T}$ should be a straight line, with $A\sqrt{\Gamma}$ as the intercept and B as the slope. As suggested by Eqs. (5.7) and (5.8), the slope B should be essentially independent of the oxygen content of the lattice (and thus independent of the amount of Y_2O_3 added, as long as it is not too large). Also, the square of the intercept divided by the slope should be proportional to the scattering cross section, Γ. That is, we expect that the ratio

$$\Gamma \propto \left(\frac{\text{intercept}}{\text{slope}}\right)^2 \qquad (5.14)$$

provides a measure of the scattering cross section. (Actually, the square of the intercept itself is proportional to Γ. The reason for choosing the ratio of intercept to the slope is to minimize the possible effects of oxygen content on other factors in A and B from Eqs. (5.7) and (5.8).) Data given by Jackson et al. are plotted as described in Figs. 5.4(a) through (d) for samples of AlN made using between 1 wt% and 8 wt% Y_2O_3 added and sintered for 100 minutes at 1850°C [12]. Note that the data can be adequately represented by straight lines consistent with Eq. (5.10). Also note that the slopes for samples with 2%, 4%, and 8% Y_2O_3 are virtually identical, consistent with expectations, although that for the sample with 1% Y_2O_3 is a factor of 2 off. The corresponding intercepts, slopes and (intercept/slope)^2 are listed in Table 5.1. The (intercept/slope)^2 ranges between ~5,425 for a sample with 1% Y_2O_3 added to ~6.2 for a sample with 8 wt% Y_2O_3 added. Because $\Gamma \propto x_s$, where x_s is the concentration of dissolved oxygen in the AlN lattice, it is seen that in the sample with 8 wt% Y_2O_3, the oxygen content is 6.2/5425 or about 0.0011 of that in the sample containing 1 wt% Y_2O_3. That is, addition of 8 wt% Y_2O_3 has been very effective in purifying the AlN lattice. Based on the oxygen content of the AlN powder used for this work, samples with 1 wt% Y_2O_3 correspond to the formation of $\mathbf{Y_3A_5}$ as the aluminate phase and the composition could correspond to the three-phase field comprising AlN–ALON–$\mathbf{Y_3A_5}$, signifying little purification of the AlN lattice compared to AlN without any Y_2O_3 added. By contrast, samples with 8% Y_2O_3 correspond to the formation of $\mathbf{Y_2A}$ as the aluminate phase. This once again shows the profound role of thermodynamic effects, namely that the activity of Al_2O_3 in AlN when $\mathbf{Y_3A_5}$ (or lower $\mathbf{Y}$) is present is orders of magnitude higher than when $\mathbf{Y_2A}$ is present. The consequence of this thermodynamic effect is that the thermal conductivity of AlN corresponding to this phase field is rather low.

5.3.3 Kinetics of Oxygen Removal and Microstructural Changes

We have so far discussed the role of thermodynamics in scavenging oxygen out of AlN grains and forming aluminate phases along the grain boundaries. The aluminate phases have very low thermal conductivities, and thus their net volume fraction should be as small as possible. Note that as the amount of the additive is increased, there are two opposing factors that come into

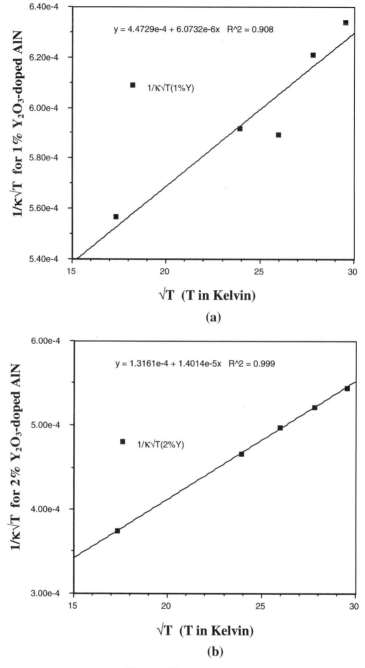

Fig. 5.4. (a) A plot of $1/\kappa\sqrt{T}$ vs. $\sqrt{T}$ for a sample sintered with 1 wt% Y_2O_3; (b) a plot of $1/\kappa\sqrt{T}$ vs. $\sqrt{T}$ for a sample sintered with 2 wt% Y_2O_3; (c) a plot of $1/\kappa\sqrt{T}$ vs. $\sqrt{T}$ for a sample sintered with 4 wt% Y_2O_3; (d) a plot of $1/\kappa\sqrt{T}$ vs. $\sqrt{T}$ for a sample sintered with 8 wt% Y_2O_3.

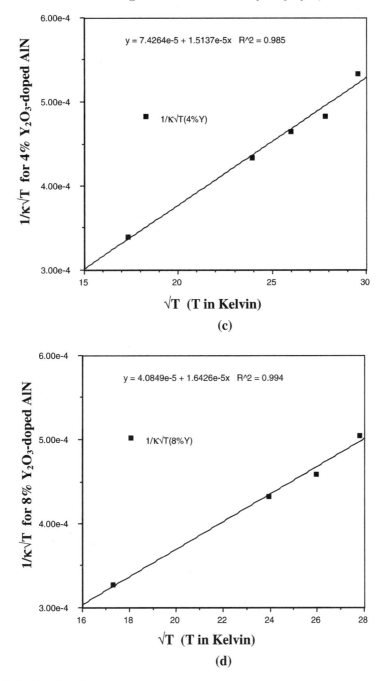

(c)

(d)

Fig. 5.4. Continued

Table 5.1.

Y_2O_3 Content (%)	Intercept $A\sqrt{\Gamma}$	Slope B	$\left(\dfrac{\text{intercept}}{\text{slope}}\right)^2$
1	4.473×10^{-4}	6.073×10^{-6}	$\sim 5,425$
2	1.316×10^{-4}	1.401×10^{-5}	~ 88
4	7.426×10^{-5}	1.514×10^{-5}	~ 24
8	4.085×10^{-5}	1.643×10^{-5}	~ 6.2

play: (1) An increase in the additive content lowers the $a_{Al_2O_3}$, increases the phonon mean free path, and thus increases thermal conductivity; and (2) an increase in the additive content increases the volume fraction of the low-thermal-conductivity aluminate phases and thus decreases the overall conductivity. It is thus expected that the thermal conductivity would initially increase rapidly with increasing additive content, reach a maximum, and thereafter exhibit a slow decrease in thermal conductivity. This has indeed been observed. Figure 5.5 shows thermal conductivity as a function of the volume fraction of secondary oxide phases for Y_2O_3 as the additive. Similar observations have been made with Sm_2O_3 and Er_2O_3 as additives, and similar observations are expected with other oxygen-scavenging oxides. The maximum in κ approximately corresponds to the maximum level of purification, and the corresponding three-phase field is expected to contain "purified" AlN,

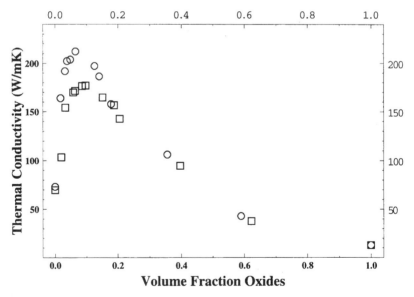

Fig. 5.5. Thermal conductivity as a function of volume fraction of the yttrium aluminate phases in AlN–Y_2O_3 compositions sinter/annealed at 1850°C. Squares = 100 minutes, circles = 1000 minutes.

the aluminate phase with stoichiometrically the largest amount of the additive, and the additive itself. That is, with Y_2O_3 as the additive, the three-phase field corresponds to AlN, $\mathbf{Y_2A}$, and $\mathbf{Y}$; or it could correspond to the existence of AlN, YN, and $\mathbf{Y}$ (Y_2O_3). Beyond the maximum, a further increase in the additive merely increases the amount of the oxide phases, without causing further purification of the AlN lattice.

The dependence of κ on the volume fraction of secondary phases can be readily described in terms of rules of mixtures. Numerous phenomenological models have been described in the literature, which include simple series and parallel models, as well as models that take into account dispersion of one phase into another [14], [15], [16]. Several articles have discussed this at length; it will not be discussed further in this chapter. Also, from a practical standpoint, AlN-containing materials beyond the maximum are of little interest.

Figure 5.5 also shows thermal conductivity as a function of the volume fraction of oxide phases for two different thermal treatments: (1) Sinter/annealed at 1850°C for 100 minutes; and (2) sinter/annealed at 1850°C for 1000 minutes. An important point to note is that for small volume fractions of the oxide phases, the thermal conductivity of samples sinter/annealed for 1000 minutes is considerably greater than that for samples sinter/annealed for 100 minutes. Figure 5.6 shows thermal conductivity as a function of

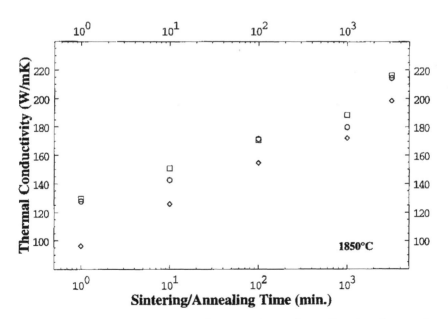

Fig. 5.6. Thermal conductivity as a function of sinter/annealing time for samples of AlN containing various rare earth oxides (Ln_2O_3): Y_2O_3 (squares), Er_2O_3 (circles) and Pr_2O_3 (or Pr_6O_{11}) (rhombus). The amount of Ln_2O_3 added was such that the molar ratio of Ln_2O_3 to Al_2O_3 (as determined by chemical analysis of the starting powder) was one.

sinter/annealing time at 1850°C for samples containing Y_2O_3, Er_2O_3, or Pr_2O_3 (Pr_6O_{11}). The amount of the rare earth oxide added was in an equimolar proportion to the actual Al_2O_3 present in the powder, as determined by chemical analysis. Densification in these systems is known to occur by liquid phase sintering, which is essentially complete within a few (<10) minutes at temperature. Thus, most of the time is spent in annealing the samples. Based on thermal conductivity measurements, it is evident that significant changes must be occurring beyond the original densification stage. These changes are at both the submicrostructural level and the microstructural level.

At the submicrostructural level, the main change that occurs is the removal of oxygen (more accurately Al_2O_3) from within the AlN grains to the grain boundaries, where the oxide additive forms a liquid phase with surface Al_2O_3 present. The removal of dissolved oxygen (more precisely Al_2O_3) can in principle be achieved by either a dissolution-reprecipitation process comprising dissolution of the original impure (oxygen-rich) AlN grains and reprecipitation of pure AlN grains or by solid-state diffusion of oxygen (that of dissolved Al_2O_3) from within the grains to the grain boundaries. While dissolution-reprecipitation is a possible mechanism (that can occur in parallel), we do not discuss it further as little relevant data are currently available to make even a semiquantitative estimate. This mechanism is nevertheless most certainly operative, at least during the subsequent annealing stage, as will be discussed shortly. For now, we will restrict our discussion to the removal of oxygen (Al_2O_3) by solid-state diffusion from within the grains to the grain boundaries. As the activity of Al_2O_3, $a_{Al_2O_3}$, at the grain boundaries is very low (due to the presence of the additive or additive-rich aluminates), there is a thermodynamic driving force for the removal of Al_2O_3 from within the AlN grains to grain boundaries. Because the concentration of dissolved Al_2O_3 is rather small, the use of simple Fick's laws is reasonable. Diffusion of Al_2O_3 in AlN is expected to occur in such a way that Al diffuses on the Al-sublattice while O diffuses on the N-sublattice. Virkar and coworkers have analyzed the data on the thermal conductivity of AlN as a function of annealing time assuming diffusion as the dominating kinetic process [13]. A convenient method of studying the kinetics of oxygen removal is to measure the thermal conductivity at room temperature as a function of time of annealing at the sinter/annealing temperature. The kinetics of conductivity (measured at room temperature) change with time (at the sinter/annealing temperature) can be adequately described by an Avrami-type equation as follows

$$X(t) = \frac{\kappa(t) - \kappa(0)}{\kappa(\infty) - \kappa(0)} = 1 - \exp\left[-\left(\frac{t}{\tau}\right)^m\right], \tag{5.15}$$

where $\kappa(0)$ is the initial thermal conductivity immediately after the sintering stage, $\kappa(\infty)$ is the thermal conductivity after long-term annealing (a stable value), $\kappa(t)$ is the thermal conductivity after annealing for time t, τ is the requisite time constant for the underlying kinetic process, and $X(t)$ is the normalized conductivity function, which is a measure of the extent to which purification has occurred. The range of $X(t)$ is from 0 to 1. The time constant for the process, τ, is related to the grain size, d, and the chemical diffusion coefficient (of Al_2O_3 in AlN), $\tilde{D}$, by an equation of the type

$$\tau \approx a\frac{d^2}{\tilde{D}}, \tag{5.16}$$

where a is a positive constant, on the order of unity. The experimental data at 1850°C can be adequately described by Eq. (5.15) with a τ on the order of 100 minutes assuming a grain size of ~5 microns. This corresponds to a $\tilde{D}$ of about 10^{-11} cm^2/s. No information is available on the diffusion of Al_2O_3 in AlN. However, given the fact that diffusion is notoriously sluggish in these covalently bonded materials and the fact that when Al_2O_3 is dissolved in AlN, transport on the N-sublattice is further suppressed and dictates the overall diffusion of Al_2O_3 (which is the chemical diffusion coefficient of Al_2O_3) in AlN (assuming diffusion to occur by a vacancy mechanism). The preceding value of $\tilde{D}$ is only approximate as many other processes occur simultaneously, such as grain growth and dissolution-reprecipitation. Thus, the measured time constant, τ, embodies many phenomena in addition to diffusion, and at the present time there appears to be no simple way of separating one from the other.

5.3.4 Long-Term Annealing and Microstructural Changes

Many significant and important changes occur during the sinter/annealing processes, which have a profound effect on the thermal conductivity of sintered polycrystalline AlN with oxide additives. From the standpoint of sintering, the most significant aspect is the formation of a liquid phase, which facilitates densification. It is well known that in most liquid-phase sintered materials, densification occurs rapidly—within minutes, or even seconds. AlN is no exception to this general observation. Rapid densification requires that the liquid phase completely wet the grains of the host material, here AlN, for full densification to occur. Thus, one concludes that at least in the initial stages, the oxide-rich phase must wet AlN grains. The typical volume fraction of the oxide phase in a sintered sample with κ on the order of 180 W/mK is about 0.1. The oxide phase has a κ of about 5 W/mK. If the oxide phase were to continue to completely wet AlN grains after processing is complete, the pertinent thermal conductivity model for the two-phase materials would be the series model (because the high-thermal-conductivity AlN grains would then be isolated from each other, surrounded by the oxide phase), given by

$$\frac{1}{\kappa} = \frac{1 - V_v}{\kappa^o_{AlN}} + \frac{V_v}{\kappa_{oxide}}, \tag{5.17}$$

where κ^o_{AlN} is the thermal conductivity of pure AlN (~320 W/mK) and V_v is the volume fraction of the oxide phase. For these values, the estimated κ is

$\sim$44 W/mK, which is much lower than the observed value. This clearly implies that the oxide phase does not completely coat AlN grains after sintering is nearly complete and during the subsequent annealing stage, and thus it also does not completely coat AlN grains in the final material cooled to room temperature. The other extreme limit is the parallel model, in which both phases form a completely contiguous network with a high degree of intraphase contiguity. According to this model, the thermal conductivity is given by

$$\kappa = (1 - V_v)\kappa_{\text{AlN}}^o + V_v\kappa_{\text{oxide}}. \tag{5.18}$$

The estimated κ according to the parallel model is $\sim$288 W/mK. The experimentally measured value is on the order of 180 W/mK, showing that it lies between the two extreme limits. As stated earlier, there are numerous phenomenological equations, the results of which lie between these two limits [14], [15], [16].

From the standpoint of this discussion, the most significant point is that the thermal conductivity of samples even after a rather short sinter/annealing time is much greater than can be described by the series model, which implies that the grain boundary oxide phase must undergo dewetting soon after the sintering stage is complete. Indeed, it has been observed using transmission electron microscopy (TEM) that the oxide phase does not completely wet the grain boundaries even after sintering for a very short time (1 minute). (In most sintering operations, the actual time a sample spends at elevated temperatures may be greater by several minutes due to the thermal inertia of the furnace.) Figure 5.7 shows a TEM micrograph of a sample containing

Tokuyama Soda AlN - 4.9 wt% Y_2O_3

Fig. 5.7. Transmission electron micrographs (TEM) showing convex-shaped aluminate particles along boundaries of AlN grains. The dihedral angle is greater than $72.54°$.

4.9 wt% Y_2O_3 sintered for 1 minute (hold time at temperature) at 1850°C. It is readily seen that the second phase does not completely coat AlN grains but exists as isolated particles. What is even more significant is that the secondary oxide phase along three-grain junctions exhibits a convex surface, indicating that the dihedral angle is greater than 60°. Measurements have shown that in fact the dihedral angle is greater than 72.54°, the latter value being the included angle between the intersecting faces of a regular tetrahedron. The significance of this statement is as follows. In a polycrystalline material with an equiaxed grain structure, the grain shape is usually a regular tetrakaidecahedron (a regular polyhedron with fourteen sides, which can be created by cutting cube corners along {111} type faces). Four such grains meet at a corner, where the four three-grain edges meet. These four three-grain edges are equivalent to joining the center of a regular tetrahedron to the four corners. A second phase precipitate at a four-grain junction (corner) exists in the form of a tetrahedron. If the dihedral angle is less than 72.54°, the faces of the second-phase tetrahedron are concave outward, and if the dihedral angle is greater than 72.54°, the faces of the tetrahedron are convex outward. If the dihedral angle is thus greater than 72.54°, the second phase will always exhibit positive curvature inside the polycrystalline body and thus would have greater chemical potential inside the polycrystalline material than if it existed as a (nearly) flat bulky particle completely out of the polycrystalline material. In such a case, the second phase is thermodynamically unstable in the polycrystalline body—and given enough time at temperatures high enough for kinetics to be fast enough, it will have a tendency to migrate to the surface of the body.

The observation that the dihedral angle is greater than 72.54° has profound implications from the standpoint of processing of high-thermal-conductivity AlN ceramics. This is because it suggests the possibility that the oxide volume fraction of a sintered AlN polycrystalline material can in principle be reduced by simply annealing for a long enough time that the second phase migrates to the surface. This indeed has been observed in samples annealed for long periods of time. Figure 5.8(a) shows the near surface regions of a sample of AlN with 4.9 wt% Y_2O_3 added that was sinter/annealed at 1850°C for more than 3000 minutes. The light region is the yttrium-aluminate phase, which has migrated from the interior of the same sample to the surface. Figure 5.8(b) shows the interior of the sample. As is clearly seen, the volume fraction of the aluminate phase is very small in the interior, as most of it has migrated to the surface. The process of migration may occur by solid-state diffusion or by dissolution-reprecipitation. Regardless of the actual mechanism involved, it is clear that the thermodynamic driving force for its expulsion from the bulk is related to surface energy (wetting characteristics) considerations. Another important point requiring emphasis is that regardless of whether the second phase is liquid or solid, the kinetics of expulsion are expected to be limited by rearrangement of the bulk AlN grains and thus is expected to be sluggish. Further, because transport involves macroscopic distances, the thicker the sample, the greater the time required for migration to the surface.

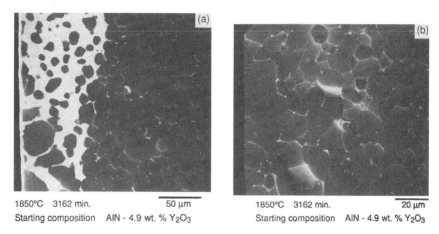

1850°C   3162 min.   50 µm
Starting composition   AlN - 4.9 wt. % Y₂O₃

1850°C   3162 min.   20 µm
Starting composition   AlN - 4.9 wt. % Y₂O₃

Fig. 5.8. Scanning electron micrographs (SEM) of AlN-4.9 wt% Y_2O_3 sinter/annealed at 1850°C for more than 3000 minutes: (a) edge of the sample showing migration of the aluminate phase toward the surface; (b) interior of the sample.

Migration of the oxide phase to the surface thus further improves the thermal conductivity of AlN, because the volume fraction of the low-thermal-conductivity oxide phase is effectively lowered in the bulk. To ensure that the thermal conductivity of the entire sample is enhanced, it is necessary that the oxide phase migrated to the surface must be somehow removed. It is known that at the elevated temperatures required for the processing of AlN under certain conditions, evaporation of the oxide phase from the surface can be facilitated. This can be achieved by suitably adjusting the atmosphere. Another option consists of simply grinding away the surface layer after the material has been cooled to room temperature, which is readily done for components of simple shapes.

5.4 Summary

High-purity aluminum nitride, AlN, is an excellent electrical insulator and an intrinsically high-thermal-conductivity material, with κ greater than $270\,\mathrm{W/mK}$ at room temperature, and has potential applications in the microelectronics industry, as well as a heat sink in many other applications. The dominant contribution to thermal conduction in AlN is due to phonons. AlN usually contains a small amount of oxygen in the form of Al_2O_3 dissolved in its lattice. The presence of dissolved Al_2O_3 has a detrimental effect on the thermal conductivity of AlN, because dissolved Al_2O_3 creates aluminum vacancies, V_{Al}, which scatter phonons and effectively lower its thermal conductivity. To realize high thermal conductivity, it is necessary that the oxygen concentration from the AlN lattice is reduced to as low a value as possible. AlN is also a highly refractory material with covalent bonding, which makes densification by solid-state sintering difficult, if not impossible. Dense samples containing AlN can be made by introducing a small amount of rare earth or alkaline earth oxide into AlN powder, forming a powder compact, and heating to a high temperature ($\sim$1850°C) in a reducing atmosphere. Introduction of the oxides not only facilitates sintering by a liquid-phase mechanism, but also serves to purify the AlN lattice by scavenging Al_2O_3 in the form of aluminates. The kinetics of sintering is generally very rapid due to the presence of a liquid phase. The ability of a given oxide additive to scavenge Al_2O_3 depends on the free energy of formation of respective aluminates. The more stable the aluminate relative to the individual oxides, the lower the equilibrium activity of Al_2O_3 dissolved in AlN, the greater the degree to which AlN can be purified, and the higher the thermal conductivity. Experimental work shows that thermodynamics and phase equilibria determine the degree to which purification of the AlN lattice can be achieved. After the densification of AlN by liquid-phase sintering is complete, continued annealing leads to further improvement in thermal conductivity, which arises due to two phenomena. The first is the removal of Al_2O_3 from within the grains to grain boundaries in the form of aluminates, which can occur by solid-state diffusion or dissolution-reprecipitation or both. The second phenomenon is the removal of the aluminates to the surface of the sample, which can also occur by solid-state diffusion or dissolution-reprecipitation or both. The driving force for the expulsion of the second phase to the surface of the sample is related to surface energy considerations, because the second phase dewets, as evidenced by the observation that the dihedral angle is greater than 72.54°. Fabrication of high-thermal-conductivity AlN ceramics starting with an impure AlN powder is an excellent textbook example wherein basic concepts in solid-state physics, such as lattice conduction of thermal energy, basic concepts in thermodynamics and kinetics of condensed phases, and traditional methods of ceramic processing, can be put to use.

References

[1] M. P. Borom, G. A. Slack, and J. W. Szymaszek, *Amer. Ceram. Bull.* **51**, 852–6 (1972).

[2] G. A. Slack, *J. Phys. Chem. Solids* **34**, 321–35 (1973).

[3] G. A. Slack, R. A. Tanzilli, R. O. Pohl, and J. W. Van der Sande, *J. Phys. Chem. Solids* **48**, 641–7 (1987).

[4] C. Kittel, *Introduction to Solid State Physics*, third edition, Wiley, New York (1967).

[5] B. Abeles, *Phys. Rev.* **131**, 1906–11 (1963).

[6] G. A. Slack, *Phys. Rev.* **126**, 427–41 (1962).

[7] K. Komeya, H. Inoue, and A. Tsuge, *Yogyo-Kyokaishi* **89**, 330–6 (1981).

[8] K. Komeya, A. Tsuge, and H. Inoue, U. S. Patent No. 4435513, Mar. 1984.

[9] N. Kuramoto, H. Taniguchi, and I. Aso, *Amer. Ceram. Soc. Bull.* **68**, 883–7 (1989).

[10] S. Prochazka and C. F. Bobik, in *Materials Science Research* **13**, ed. G. C. Kuczynski, Plenum Press, New York and London, 321–32 (1979).

[11] Y. Kurokawa, K. Utsumi, H. Takamizawa, T. Kamata, and S. Noguchi, *IEEE Trans. Compon. Hybrids, Manuf. Technol.* **CHMT-8**, 247–52 (1985).

[12] T. B. Jackson, A. V. Virkar, K. L. More, R. B. Dinwiddie, Jr., and R. A. Cutler, *J. Amer. Ceram. Soc.* **80**, 1421–35 (1997).

[13] A. V. Virkar, T. B. Jackson, and R. A. Cutler, *J. Amer. Ceram. Soc.* **72**, 2031–42 (1989).

[14] A. Eucken, *VDI-Forschungsh.* **353**, 1–16 (1932).

[15] Z. Hashin and S. Shtrikman, *J. Mech. Phys. Solids* **11**, 127–40 (1963).

[16] D. S. McLachlan, M. Blaszkiewicz, and R. E. Newnham, *J. Amer. Ceram. Soc.* **73**, 2187–203 (1990).

6

High-Thermal-Conductivity SiC and Applications

J.S. Goela, N.E. Brese, L.E. Burns, and M.A. Pickering

6.1 Introduction

Although it does not occur in nature and was first synthesized a little more than a century ago, silicon carbide is one of the most important industrial ceramic materials, with consumption greater than one million tons per year. Discovered by Pennsylvanian Edward Acheson in 1891 and patented shortly thereafter [1], silicon carbide justified a premium price (more than $800/lb) due to its unique abrasive character. The availability of inexpensive hydro-electric power in Niagara Falls led Acheson to set up his Carborundum facility in the vicinity. Slight modifications of the original process are used today to generate hexagonal forms of SiC from the high-temperature reaction of quartz sand and petroleum coke [$SiO_2 + 3C \rightarrow SiC + 2CO$]. Silicon carbide played a key role in the industrial revolution and is still widely used as an abrasive and as a steel additive, refractory, and structural ceramic. An excellent review is available in [2].

Silicon carbide crystallizes in a close-packed structure of covalently bonded silicon and carbon atoms. These atoms are arranged so that two primary coordination tetrahedra, SiC_4 and CSi_4, where four carbon or silicon atoms are bonded to a central Si or C atom, are formed [3]. These tetrahedra are linked together through their corners and stacked to form polar structures called **polytypes**, which are alike in the two dimensions of the closed packed plane but differ in the stacking sequence in the dimension perpendicular to these planes. The stacking sequence in SiC can be described by an ABC notation. In a cubic, 3C, or β-SiC, a sequence of three planes or the ABC stacking (...ABCABC...) is repeated to form a zinc-blend structure whereas in a simple hexagonal 2H-SiC, a sequence of two planes (...ABAB...) is repeated. In addition, more than 100 polytypes of SiC exist that contain more complex stacking arrangements derived from these two forms. All these noncubic forms of SiC are known as α-SiC.

Silicon carbide (SiC) is a good material for high-heat-flux applications due to its many attractive properties, such as high thermal conductivity, which is exceeded only by diamond, low values of density and thermal expansion coefficient, and high values of hardness, elastic modules, flexural strength, and thermal shock resistance. Further, SiC is a wide-band-gap material (band gap = 2.2–2.86 eV) with good transmission in the wavelength region 0.5–6 μm, so it can also be used as windows and domes for high-speed aircraft and missiles.

The properties of SiC depend considerably on the specific method used to produce it. Four basic types of SiC are currently available: hot pressed or sintered, siliconized or reaction bonded, single crystal and chemical vapor deposited (CVD). In the hot-pressed process [4], [5], SiC powders are mixed with suitable sintering aids and grain growth inhibitors and are consolidated at high temperature and pressure to form near 100% dense SiC parts of relatively simple shapes. Use of hot isostatic pressing allows fabrication of small components of intricate shapes. Although this technique provides good mechanical properties, the resulting SiC does not provide high values of other properties such as thermal conductivity, optical transmission, or high surface quality. Further, reliable bonding techniques are required to fabricate large and complex-shaped parts.

The reaction-bonded SiC (RB-SiC) is a two-phase material consisting of SiC and 10–40% of Si [4], [5[, [6]. First, a SiC porous body is formed by casting an alpha SiC slurry and then this body is infiltrated with Si to fill the pores and yield a near 100% dense material. Thus the properties of this material are primarily determined by the Si content. This method permits fabrication and in-process repair of parts such as lightweight structures for space mirrors. However, due to the presence of particles of different thermal conductivity, refractive index, and hardness, this material does not provide very high thermal conductivity, it is not useful for transmissive optics applications, and it cannot be polished to a high degree of surface figure and finish. The last drawback is usually overcome by overcoating this material with a layer of CVD-SiC, Si or any other suitable material.

Single-crystal SiC (undoped) can provide good thermal, optical, and physical properties of interest for high-thermal-conductivity applications [7], [8]. However, single-crystal SiC is expensive, difficult to produce in large sizes, and susceptible to fracture along the cleavage planes. The hexagonal form of single-crystal SiC (α-SiC) has been produced in small sizes for semiconductor applications using sublimation and is currently available commercially. The single-crystal β-SiC, which is cubic and isotropic, is not readily available.

Chemical-vapor-deposited SiC is a superior material for high-thermal-conductivity applications [9], [10], [11], [12], [13], [14], [15], [16]. By varying the process parameters, the same CVD method can produce high-thermal-conductivity SiC with other properties such as electrical resistivity or optical transmission optimized for a variety of applications. CVD-SiC is a theoretically dense, highly pure, polycrystalline material, which is free from voids and microcracks. The CVD process permits use of near-net shape and precision replication technologies to fabricate components, which require minimal

Table 6.1. Comparison of important properties of different forms of SiC.

Material	Density (gcm^{-3})	Thermal Conductivity (Wm^{-1}K^{-1})	CTE 20–1000°C (K$^{-1} \times 10^{-6}$)	Elastic Modulus (GPa)	Polishability (Å RMS)
CVD-SiC	3.21	300–390	4.0	466	≤3
Single-crystal SiC	3.21	300–490	—	—	≤3
Reaction-bonded SiC	3.1	120–170	4.3	391	≥20
Hot-pressed SiC	3.2	50–120	4.6	451	≥50

postdeposition fabrication and polishing. Further, the CVD process is scalable. Monolithic 0.5-m-diameter or 1.0-m-long and 25-mm-thick parts have been successfully produced. This process can be further scaled to yield multimeter-size parts. Large-scale capability reduces cost and makes CVD-SiC components cost-effective in comparison to other competing materials. Finally, the CVD process is reproducible. This reproducibility has been demonstrated statistically by plotting important properties of CVD-SiC on statistical quality-control charts. A consequence of reproducibility and homogeneity of CVD-SiC is that a fabrication process can be developed to yield parts of consistent quality and finish from batch to batch.

Table 6.1 shows a comparison of important properties of different forms of SiC. We see that CVD-SiC has properties superior to all other forms of SiC except the single-crystal SiC, which is not readily available in bulk form. Specifically, CVD-SiC has substantial advantage in terms of thermal conductivity and polishability in comparison to hot-pressed and RB-SiC.

We present CVD β-SiC process and property data relevant for high-thermal-conductivity applications. Discussion has been limited to bulk CVD-SiC, and no attempt has been made to include applications of CVD-SiC coatings. In Section 6.2, the process conditions used to produce high-thermal-conductivity SiC are presented. Many applications may require not only high thermal conductivity but also optimum values of other properties such as flexural strength, optical transmission, electrical resistivity, and chemical purity. Consequently, important physical, thermal, mechanical, optical, and electrical properties of CVD-SiC are provided in Section 6.3. The high-thermal-conductivity applications of CVD-SiC are discussed in Section 6.4. Finally, a summary and conclusions are presented in Section 6.5.

6.2 CVD-SiC Process

The CVD process can be used to produce two different types of SiC—opaque and transparent. Many properties of these two forms of SiC are the same. The main difference is in the visible-infrared transmission. The transparent SiC has high transmission in the 0.5–6 μm region whereas the opaque material exhibits substantial scattering, which makes it unsuitable for those applications that require both high thermal conductivity and high transmission. Although thin samples (<0.1 mm thick) of the opaque SiC appear transparent, most transmission applications, such as windows for high-speed missiles and aircraft, require material of reasonable thickness (>1 mm). Silicon carbide has many applications where its transmission properties are not required and because transparent SiC is more difficult to produce, most of the CVD process development efforts to date have been concentrated on the opaque SiC.

Several different silicon and carbon sources have been used to produce SiC in a CVD chamber. These sources include different silanes, such as SiH_4, $SiCl_4$, and $SiHCl_3$ for Si; different hydrocarbons such as CH_4, C_3H_8, and C_5H_{12} for carbon; and carbosilanes, such as CH_3SiCl_3, $(CH_3)_2SiCl$ for both Si and carbon. Good discussions of the CVD-SiC process are provided in References [3], [14], [15], and [16]. Many studies have been performed on the pyrolysis of methyltrichlorosilane [17], [18], [19], [20], [21], [22].

In comparison, very few studies were performed to produce good-quality transparent β-SiC. Single-crystal [23] or thin-film epitaxial growth techniques [24], [25], [26] yielded good-quality β-SiC, but the size or thickness of the material produced was quite small. Postdeposition annealing of CVD β-SiC [27] had the potential to produce large areas of SiC, but the best value of the attenuation coefficient obtained at 3 μm was about 11 cm^{-1}, which is too large for high-speed missile applications. Other efforts that used CVD techniques [17], [28] and were focused specifically on improving optical transmission of β-SiC also met with limited success. Weiss and Diefendorf [17] obtained small pieces of translucent SiC (attenuation coefficient = 6 cm^{-1} at 0.6328 μm) by flowing reagents in a small slot at high speed, but when the same reagents were made to flow in a larger-area box, opaque SiC was produced. Chu and Han [28] correlated the SiC deposit morphology to infrared transmission, but they did not report specific values for the attenuation coefficient.

Opaque and transparent SiC have been produced at Rohm and Haas Advanced Materials under different process conditions described elsewhere [13], [14], [29]. The opaque SiC is currently available commercially under the trade name CVD SILICON CARBIDE®. Rohm and Haas has done limited research in producing transparent SiC [30], [31], [32], [33], [34]. This research led to preparation of higher-transmission SiC in the visible and infrared regions. To date, small samples 30 cm^2 × 1.5 mm-thick have been produced [30], [31], [32], [33], [34], [35], [36]. Additional work is required to scale the process to make it useful for many optical applications.

Figure 6.1 shows a schematic of the top and front views of the CVD-SiC deposition chamber [29]. It consists of a water-cooled stainless-steel chamber with graphite heating elements. The SiC deposition area consists of inside walls of an open deposition box. The flow of reagents is from top to bottom, although other flow directions are also possible. The carbon- and silicon-containing gas mixture reacts on the mandrel surface to produce SiC by the

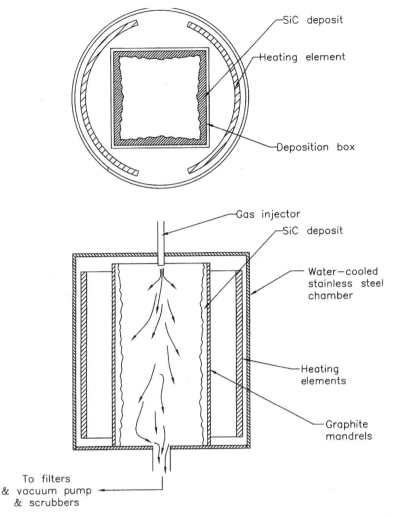

Fig. 6.1. Schematic illustration of the SiC deposition setup showing top and front views.

heterogeneous reaction. Figure 6.2 shows a picture of the CVD-SiC furnace. Using this furnace we have been able to prodice SiC parts in the meter-size range.

CVD-SiC is produced in large sheets. These sheets are cut into various sizes, and these parts are individually fabricated to the required specifications. Near-net-shape and precision-replicated parts also can be produced in the CVD process. To accomplish that, the mandrels are placed perpendicular to the flow, so that the flow impinges on them [37]. This configuration provides a more uniform SiC deposit. The thickness uniformity is further improved by rotating the mandrels at 1–2 rpm and distributing several injectors appropriately in the radial and azimuthal direction [38], [39], [40]. This arrangement

Fig. 6.2. A picture of CVD-SiC deposition chamber used to produce silicon carbide.

provides thickness uniformity of a few percent azimuthally and about 10% in the radial direction on 1-m-diameter parts.

An important concern in ceramics materials is the reproducibility of the production process to yield identical material properties from batch to batch. This process reproducibility was investigated by plotting several material properties, such as chemical purity, hardness, fracture toughness, flexural strength, grain size, thermal conductivity, and thermal expansion coefficient, on statistical quality-control charts. The analysis of these charts indicated that CVD-SiC process is highly reproducible [41].

As an example of process reproducibility, Fig. 6.3 shows the average thermal conductivity of CVD-SiC for thirty-eight different lots. For each lot, seven samples were taken from different locations of the furnace and measurements were performed by the laser flash technique at Holometrix, Bedford, MA, and also at Rohm and Haas Advanced Materials using Holometrix equipment. Because SiC samples are translucent to laser radiation at $1.06\,\mu\mathrm{m}$, they were

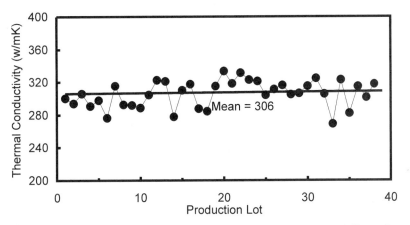

Fig. 6.3. The average thermal conductivity of CVD-SiC for 27 different lots.

made opaque by first applying a thin coating of gold by sputtering followed with spraying a thin layer of carbon on the surface. From Fig. 6.3 we see that the average thermal conductivity value is $306 \pm 16\,\mathrm{Wm^{-1}K^{-1}}$. Thus within 5%, the thermal conductivity results are reproducible in the CVD-SiC process.

6.3 Properties of CVD-SiC

CVD-SiC has been extensively characterized for important physical, mechanical, optical, thermal, and electrical properties. Table 6.2 lists important properties of opaque and transparent SiC. From Table 6.2 we see that CVD-SiC is a theoretically dense, void-free, highly pure, polycrystalline material with high oxidation, thermal shock, abrasion, and corrosion resistance [42], [43]. In addition, electrical resistivity has been tailored in the range of 0.01–1000 ohm-cm by varying the SiC process conditions with minimal effect on the other properties of CVD-SiC.

Table 6.3 shows the typical X-ray diffraction results for CVD-SiC. Also shown are relative intensity values for a SiC powder sample. The transparent SiC sample shows a very high preferred orientation along the <111> direction whereas the opaque SiC is randomly oriented. This preferred orientation may be the key to obtaining high transmission with high thermal conductivity in SiC [33].

CVD-SiC is a highly pure material. The total trace element impurity in CVD-SiC has been measured to be less than 5 parts per million by weight (ppmw) by glow discharge mass spectroscopy at Shiva Technology, Inc., Cicero, NY. Table 6.4 lists the trace element impurity concentration for fifteen important elements. Most impurities are below the detection limit of this method and fall in the fraction-of-parts-per-million range.

Table 6.2. Important properties of CVD-SiC.

Property	Average Value
Color	Dark gray (opaque SiC)
	Yellow (transparent SiC)
Crystal structure	FCC polycrystalline, β-phase
	Randomly oriented
	(opaque SiC)
	Highly oriented <111>
	(transparent SiC)
Average grain size (μm)	5–10
Transmittance, 0.6–5.6 μm	>40% (transparent SiC)
(0.5 mm thick)	
	0% (opaque SiC)
Attenuation coefficient (cm^{-1})	
@ 0.6328 μm	6.9 (transparent SiC)
	>100 (opaque SiC)
3 μm	2.2 (transparent SiC)
	>60 (opaque SiC)
Density (g cm^{-3})	3.21
Vickers hardness (1 Kg load)	2540
Fracture toughness, $K_{IC}(\mathrm{MN\,m}^{-1.5})$	2.2 (transparent SiC)
	3.1 (opaque SiC)
Elastic modulus, GPa	466
Flexural strength, MPa	470
Weibull parameters	
Modulus, m	11.45
Scale factor, MPa	462
Trace element impurities (ppmw)	<5
Thermal expansion coefficient $(10^{-6}\ \mathrm{K}^{-1})$	
@ 293K	2.2
Thermal conductivity $(\mathrm{Wm}^{-1}\mathrm{K}^{-1})$	
@ 27C	214 (transparent SiC)
	300–390 (opaque SiC)
Heat capacity $(\mathrm{Jkg}^{-1}\mathrm{K}^{-1})$	640
Electrical resistivity (ohm-cm)	4.5×10^4 (transparent SiC)
	0.01–1000 (opaque SiC)
Dielectric constant (35–50 GHz)	136
Dielectric loss (35–50 GHz)	75
Loss tangent	0.55
Refractive index	
@ 633 nm	2.635
1152 nm	2.576
1523 nm	2.566
Thermo-optic coefficient, dn/dT $(10^{-6}\ \mathrm{K}^{-1})$	
@ 2–4 μm	37

Table 6.3. X-ray diffraction data for CVD-SiC.

2θ Values (degrees)	d-spacing (Å)	Orientation	Relative Intensity (%)		
			SiC Powder Pattern	Transparent SiC	Opaque SiC
35.7	2.51	<111>	100	100	100
41.3	2.18	<200>	20	0	8.1
59.9	1.54	<220>	35	0	29.8
71.7	1.32	<311>	25	0	27.2
75.4	1.26	<222>	5	5.32	6.2
89.9	1.09	<400>	5	0	2.7
100.7	1.00	<331>	10	0	7.0
104.3	0.98	<420>	5	0	2.2
119.8	0.89	<422>	5	0	6.4

Figure 6.4 shows the grain size and microstructure of CVD-SiC. Because CVD-SiC is not readily attacked by acids and bases it is difficult to etch. Two methods that have been successful in etching SiC involve using hot KOH and fluorine plasma. In the former method, pure KOH pellets are heated to 900°C in a nickel crucible and the SiC sample is etched in the molten KOH for up to 10 minutes. In the latter method, the sample is first etched in an argon plasma for 5 minutes followed by etching in a $CF_4 + 4\%$ O_2 plasma for one hour. The plasma power used was 300 watts, frequency 13–56 MHz, and the

Table 6.4. Trace element impurity concentration as determined by gas discharge mass spectroscopy.

Element	Concentration (ppmwt)
Li	<0.001
B	0.41
S	0.03
Na	<0.01
Cl	<0.05
Ni	0.11
Co	<0.005
Cu	<0.05
Fe	0.12
Al	<0.01
Cr	<0.1
W	<0.01
Mo	<0.05
Ti	<0.005
Mn	<0.01

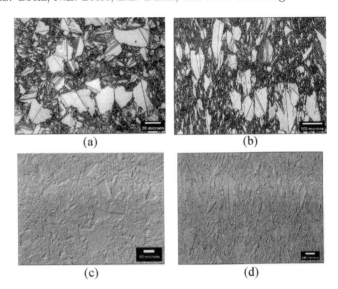

Fig. 6.4. Grain size and microstructure in CVD-SiC (a) grain structure, KOH etching, Sample 1; (b) microstructure (cross-section), KOH etching, Sample 1; (c) grain structure, plasma etching, Sample 2; (d) microstructure (cross section), plasma etching, Sample 2.

gas pressure was 350 torr. Microstructure parallel and perpendicular to the growth direction from both these methods are compared in Figs. 6.4 (a)–(d). We see that KOH etching provides clearer pictures of the grain structure. The plasma etching process does not etch the SiC surface uniformly, so only a few grains are visible. However, it does show crystallite defects as well as the columnar structure characteristic of the CVD process. SiC microstructure shows the presence of a few large grains and many small grains. The average grain size falls in the range of 5 to 10 microns. The cross section shows growth columns that have low angle deviation from the growth direction.

Transmission electron microscope images of opaque and transparent SiC samples are shown in Fig. 6.5. In a transparent SiC sample (Fig. 6.5(a)), the grains contained a low density of dislocations. Grains were about 5 to 10 microns and were almost always found to have one of their <111> direction perpendicular to the deposition surface within a few degrees [33]. In many cases, the preferred orientation was nearly "perfect" in several abutting grains. In comparison, the opaque SiC sample contained many grains and growth defects (see Fig. 6.5(b)).

The CVD-SiC samples were fractured perpendicular to the deposition surface (cross section), and the fracture patterns were examined by scanning electron microscope. Figure 6.6 shows the results. The transparent sample exhibited a highly oriented columnar structure with a deviation of only a few degrees (Fig. 6.6(a)) whereas the opaque samples did not show much preferred

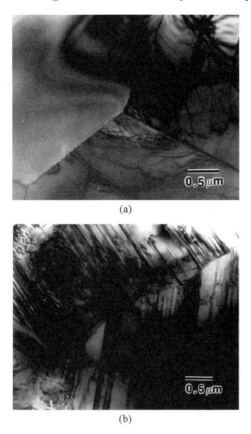

(a)

(b)

Fig. 6.5. Transmission electron microscope pictures of transparent and opaque CVD-SiC: (a) transparent SiC, (b) opaque SiC (from [33]).

orientation (Fig. 6.6(b)). These results are in accord with the X-ray diffraction data discussed earlier.

6.3.1 Thermal Properties

In this section we discuss three thermal properties—thermal conductivity, specific heat, and thermal expansion coefficient—and their variation with temperature. Also discussed is the thermal shock resistance of SiC, which is required in many high-thermal-conductivity applications.

6.3.1.1 Thermal Conductivity and Specific Heat. Thermal conductivity is the product of thermal diffusivity, density, and specific heat of the material. To obtain thermal conductivity as a function of temperature, thermal diffusivity and specific heat were measured as a function of temperature at Purdue University, Lafayette, IN, and then, from these data, thermal conductivity was

(a)

(b)

Fig. 6.6. Scanning electron microscope pictures of fracture pattern of transparent and opaque SiC: (a) transparent SiC, (b) opaque SiC (from [33]).

calculated using the density value at room temperature ($d = 3.207\,\mathrm{g\,cm^{-3}}$). Figure 6.7(a) shows the specific heat of CVD-SiC as a function of temperature in the temperature range $-150°\mathrm{C}$ to $1800°\mathrm{C}$. The specific heat values were measured using Netzsch differential scanning calorimeter up to $1200°\mathrm{C}$. Values at and above $1400°\mathrm{C}$ were extrapolated. These specific heat values are typical of polycrystalline SiC.

Figure 6.7(b) shows the thermal conductivity of CVD-SiC as a function of temperature in the range of $-150°\mathrm{C}$ to $1800°\mathrm{C}$ for two samples, which were taken from two different areas of the furnace. At $-150°\mathrm{C}$ and $1800°\mathrm{C}$, the thermal diffusivity values were not measured but extrapolated. Both samples in Fig. 6.7(b) show essentially the same thermal conductivity values indicating that CVD-SiC is homogeneous. Further, the thermal conductivity peaks at $485\,\mathrm{Wm^{-1}K^{-1}}$ near $173\,\mathrm{K}$ and the curve shows a T^{-1} dependence above $200°\mathrm{C}$.

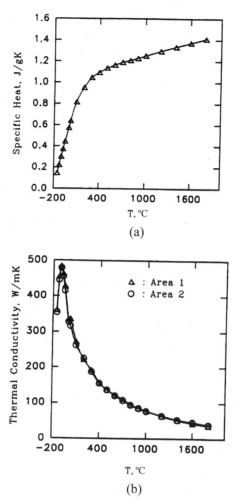

Fig. 6.7. Specific heat and thermal conductivity as a function of temperature: (a) specific heat, (b) thermal conductivity for two samples of CVD-SiC taken from two different areas of the SiC furnace.

The polycrystalline SiC produced by CVD under optimum conditions has exhibited very high-thermal-conductivity values, 300–374 $Wm^{-1}K^{-1}$ at room temperature [29]. These values are comparable to the best values that metals have exhibited (e.g., copper is 390 $Wm^{-1}K^{-1}$) and are exceeded only by diamond. Silicon-carbide conductivity depends on CVD process conditions such as substrate temperature, furnace pressure, and flow rates of reagents.

The mean value of thermal conductivity for 78 production runs came out to about 300 $Wm^{-1}K^{-1}$ with a range of about 374–120 $Wm^{-1}K^{-1}$ [29]. The conductivity values less than 200 $Wm^{-1}K^{-1}$ are usually obtained from samples taken from near the plate edges where the flow is stagnant. Higher conductivity values are produced at those locations in the reactor where optimum process conditions exist.

The variation of thermal conductivity along the growth direction in CVD-SiC is shown in Fig. 6.8. A SiC sample about 25 mm thick was taken and sliced into seven samples, each 2 mm thick. The conductivity was measured along the growth direction using a laser flash technique, and the data are shown in Fig. 6.8. We see that the thermal conductivity for all samples fall into a narrow range between 327 and 363 $Wm^{-1}K^{-1}$. Thus the conductivity variation is only ±5% around the mean value, 345 $Wm^{-1}K^{-1}$. This variability is much less than the variability observed in other CVD materials, such as diamond.

Variations across columnar grains are also interesting. The thermal conductivity has a higher value perpendicular to the deposition surface (i.e., along the columnar grains) than parallel to it. This is typical of CVD materials because in the latter case the transport of heat is affected by the presence of grain boundaries. The conductivity difference in the two directions is about 15%.

The phonon mean free path in SiC is on the order of tens of nanometers. Consequently, the thermal conductivity of CVD-SiC can be affected by crystal imperfections such as stacking faults. In order to assess this effect, we analyzed X-ray diffraction scans of CVD-SiC. Typically the X-ray diffraction patterns collected from cubic β-SiC ($a = 4.35$ Å) are expected to have sharp diffraction (Bragg) peaks at the following locations in 2θ, if using Cu Kα radiation: $35.8°$ {111} and $41.5°$ {200}. Stacking fault features appear as broad features, several degrees in width, near their Bragg counterparts [44], [45], [46].

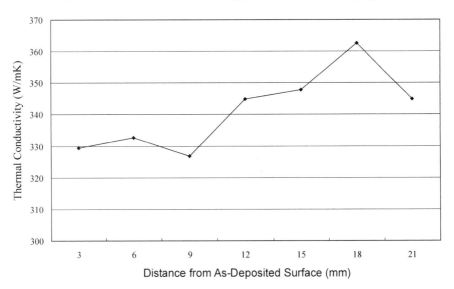

Fig. 6.8. Variation of CVD-SiC thermal conductivity along the thickness (growth direction).

To monitor the stacking faults in CVD-SiC, we sum the total diffraction intensity from 32° to 48° and subtract the baseline and the intensity attributable to the Bragg diffraction (35.5° to 36° and 41.25° to 41.75°). The ratio of the remaining intensity to the total intensity is a measure of the amount of irregularity in the crystalline order. Twenty-eight samples were taken from different SiC production runs and the X-ray intensities attributable to stacking faults were determined for each sample and plotted as shown in Fig. 6.9. The thermal resistance (inverse thermal conductivity) is observed to vary roughly linearly with the stacking fault density, as would be expected if the added thermal resistance is due to the presence of stacking faults. We also note that the thermal resistivity extrapolated to zero stacking fault density corresponds to a thermal conductivity of 500 W/mK, in agreement with the value for single-crystal SiC.

It has been shown previously that increasing the deposition temperature in the CVD-SiC process increases the material's thermal conductivity [47]. This effect may be due to a reduction in stacking faults in the CVD-SiC rather than an increase in grain size or it may be attributable to a combination of the two.

6.3.1.2 Thermal Expansion Coefficient. Figure 6.10 shows the thermal expansion coefficient (CTE) of CVD-SiC as a function of temperature in the range of

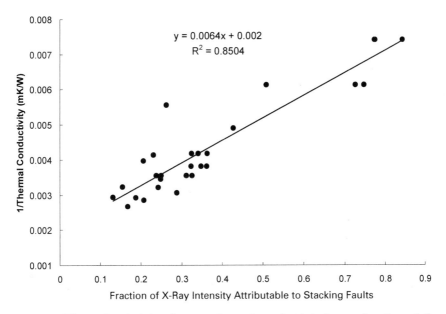

Fig. 6.9. Thermal resistivity (inverse thermal conductivity) as a function of the stacking fault density. To calculate the stacking fault contribution, we summed the area from 32 to 48 degrees two theta (Cu Ka), subtracted the baseline (average of 30–2 and 55–6°), then subtracted the Bragg peak areas (35.5–36.0 and 41.25–41.75), and finally normalized stacking fault area to total area.

133 K to 1273 K. These measurements were made using a differential dilatometer supplied by Theta Industries, Inc. We see that CTE decreases rapidly at low temperatures and remains relatively constant at high temperature. This behavior is typical of crystalline materials such as CVD-SiC.

6.3.1.3 Thermal Shock Resistance. Many high-thermal-conductivity applications where high heat loads are present require that the material exhibits good resistance to thermal shock. The thermal shock parameter, R, is defined as $\sigma\kappa\,(1-\nu)/\alpha E$, where σ is the flexural strength, κ is the thermal conductivity, ν is the Poisson ratio, α is the thermal expansion coefficient, and E is the elastic modulus. This parameter provides a relative indication of thermal shock resistance of materials when the Biot number, $hL/\kappa \leq 1$ [48]. Here h is the heat transfer coefficient and L is the characteristic dimension, which could be the thickness of the window or dome. Table 6.5 compares the thermal shock parameter of CVD-SiC with that of several competing materials [48], [49]. We see that CVD-SiC provides a thermal shock parameter value that is significantly greater than those of all other materials, except CVD diamond. However, diamond is a relatively expensive material and is difficult to polish and obtain in large sizes. Further CVD diamond grown in thick layers exhibits growth in grain size with increasing thickness that result in considerable thermal conductivity variation along the thickness of the material. Overall, CVD-SiC is a preferred material for high thermal shock resistance.

6.3.2 Mechanical Properties

In this section we discuss two mechanical properties—flexural strength and elastic modulus—and their variation with temperature.

Table 6.5. Comparison of thermal shock resistance of some important optical materials.

Material	Flexural Strength σ (MPa)	Elastic Modulus E (GPa)	Poisson Ratio ν	Thermal Conductivity κ (Wm^{-1}K^{-1})	CTE α (10^{-6} K^{-1})	Thermal Shock Parameter R
CVD-SiC	420	466	0.21	300	2.2	97
Sapphire	400	380	0.27	24	8.8	2.1
Spinel	160–190	190	0.26	14.6	8.0	1.2–1.39
ALON	300	315	0.24	12.6	7.8	1.02
Yttria	116	164	0.3	14	7.1	0.94
MgF$_2$	100	115	0.3	16	11.0	0.89
CVD diamond	160–400	1140	0.069	$\leq$2300	0.8	$\leq$939
GaP	100	103	0.31	97	6	10.8
GaAs	60	86	0.31	53	6	4.3
CVD-ZnS	103	75	0.29	16.7	7	2.3

Figure 6.11 shows the flexural strength and elastic modulus as a function of temperature. The flexural strength measurements were made at the University of Dayton Research Institute in the temperature range of 79 K to 1723 K using an Instron machine. For these data, all the beams were prepared with a surface finish of ~0.5 μm RMS. The solid line is a least square linear regression fit to the data points. Error bars represent standard deviation in the measured data. We see that CVD-SiC strength increases with temperature (Fig. 6.11(a)). This effect has been observed previously for CVD-SiC and is attributed to a small plastic deformation that occurs at crack tips at higher temperatures.

The elastic modulus was measured at the University of Dayton Research Institute using a Grindo Sonic, MK3 (J.W. Lemmens Co). From Fig. 6.11(b) we see that the sonic modulus decreases by only 10% when the temperature increases from room temperature to 1500°C.

Because SiC is a brittle ceramic material, it is susceptible to the flaw-induced fracture. The flaw size, C_f in a brittle material is given by the formula [14]:

$$C_f = 0.79(K_{1C}/\sigma)^2, \tag{6.1}$$

where K_{1C} is the fracture toughness and σ is the strength of the SiC part. Because the fracture toughness of the material is a constant, the strength of the part depends on the size of the flaw in the material, which in turn depends on the volume of the material used in the part. Thus the larger the part, the higher the probability of finding a larger flaw. For SiC with $K_{1C} = 3.4\,\text{MNm}^{-1.5}$ and $\sigma = 421\,\text{MPa}$, the flaw size is about $52\,\mu\text{m}$, which is quite small and is a few times the grain size of the material. The maximum

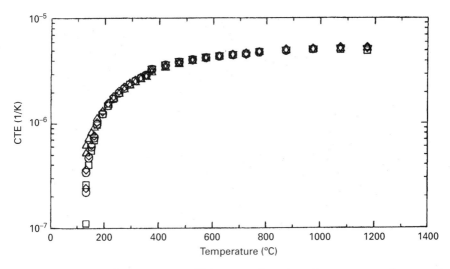

Fig. 6.10. Thermal expansion coefficient as a function of temperature for several samples of CVD-SiC.

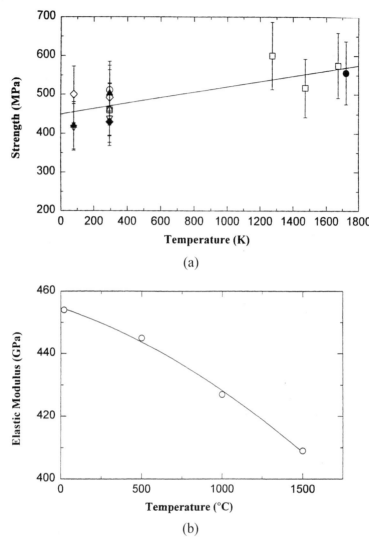

Fig. 6.11. Flexural strength and elastic modulus as a function of temperature for CVD-SiC: (a) flexural strength, (b) elastic modulus.

allowable stress, σ, in large parts can be calculated from the following formula:

$$\sigma = \sigma_1 (A_1/A)^{1/m}, \tag{6.2}$$

where σ_1 is the mean fracture stress for the test specimens, A is the area of the large part, A_1 is the area of the test specimen, and m is the Weibull modulus. For SiC, $m = 11.45$, $\sigma_1 = 421\,\mathrm{MPa}$, and $A_1 = 160\,\mathrm{mm}^2$. Consequently, for a 1-m-diameter part, the allowable maximum stress in the part is $200\,\mathrm{MPa}$. For as-grown SiC surfaces, however, the value of m was determined to be about 4 with $\sigma_1 = 262\,\text{MPa}$. In this case the allowable maximum stress in the 1-m-diameter part is only about 31 MPa, which is quite small. These calculations indicate that while producing large parts by the CVD process, SiC deposits should not be stressed beyond the allowable values during furnace cooldown.

The flexural strength of the CVD-SiC can be enhanced by annealing, appropriate surface treatment, or introducing a compressive stress in the material. The flexural strength of CVD-SiC increased by 37% to 53% when it was heated in the temperature range of 600°C to 1000°C and quenched in water or annealed in flowing N_2. In the case of thermal shock by water quenching, the increase started at 600°C and became relatively constant above 800°C. This increase was attributed to an increase in compressive residual stresses and partially due to healing of machining flaws [50].

6.3.3 Electrical Properties

Electrical resistivity is another property that may require tailoring when SiC is used for high-thermal-conductivity applications. The electrical resistivity of CVD-SiC has been tailored in the range of 0.01–1000 ohm-cm by varying the CVD process conditions with minimal effect on the other properties of CVD-SiC, particularly thermal conductivity. When SiC is produced by the opaque CVD-SiC process, SiC with an electrical resistivity in the range of 1–100 ohm-cm is produced. This variation in electrical resistivity may be due to a small amount of contaminants, which are invariably present in the process but escape detection by the gas-discharge mass spectroscopic analysis. SiC with resistivity greater than 100 ohm-cm was produced by controlling this contaminant "noise" level in the CVD-SiC process. Material with resistivity less than 1 ohm-cm was produced by introducing appropriate concentration of dopants in the CVD-SiC process. As higher concentration of dopants were introduced, the process yielded lower resistivity values and less spread in the data. Even at very high dopant concentrations, the thermal conductivity of CVD-SiC was minimally effected.

Figure 6.12 shows plots of high- and low-resistivity samples as a function of sample temperature in the temperature range of 20°C to 500°C. We see that the resistivity of SiC decreases as its temperature increases. This effect is typical of SiC and is due to increase in carrier concentration as the sample temperature increases. The SiC samples were held at 500°C for 500 hours. No hysteresis effects were observed, and the resistivity values as function of temperature remained the same.

6.3.4 Optical Properties

Chemical-vapor-deposited SiC has been polished to the smoothest surfaces ever produced. The surface roughness has been measured to be less than 0.5 Å RMS by the Zygo heterodyne profiler, less than 1 Å RMS by the Talystep

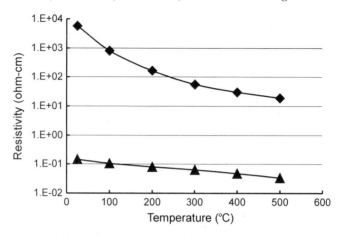

Fig. 6.12. High- and low-resistivity SiC as a function of temperature.

mechanical contact profiler, and less than 5 Å RMS by the atomic force microscope. These measurements were made at the Naval Weapons Center, China
Lake, CA. The total integrated scatter was measured to be less than or equal
to 1×10^{-4} at $0.6328 \, \mu m$ at the Naval Weapons Center, China Lake, CA and
TMA (now Schmidt, Portland, OR). The bidirectional reflection distribution
function (BRDF) was measured to be less than or equal to 5×10^{-6} from 10 to
80 degrees from the specular at $0.6328 \, \mu m$.

The reflectivity of CVD-SiC in the visible-infrared regions is less than
20%. Furthermore CVD-SiC exhibits anomalous scattering in the infrared
region [51], that is, it does not follow the λ^{-4} scattering law. At ± 10 degrees
from specular, BRDF was measured to be less than 1×10^{-5} at $0.325 \, \mu m$ and
$0.6328 \, \mu m$, but 3×10^{-4} at $10.6 \, \mu m$. Such a high value of BRDF at $10.6 \, \mu m$
indicates that the scattering is not occurring topographically from the surface
according to λ^{-4} law. The reflectivity of CVD-SiC was substantially increased
by applying a coating of protective silver and gold. A silver-coating thickness
of at least 500 Å is required on CVD-SiC to eliminate the anamolous scattering
effect and obtain good broadband reflectance in the visible-infrared region.

Because gold does not adhere well to SiC, a bonding layer of chrome/nickel
or other reactive metal is usually applied first by the vapor-deposition process. Subsequently, coating methods such as electroplating, evaporation, and
magnetron sputtering are used to apply the gold coatings. Electroplated gold
coating of thickness 1000–2500 Å have been successfully applied on the polished surface of CVD-SiC. This coating has yielded a reflectivity of less than
98% at $0.8 \, \mu m$ with a surface roughness of less than 5 Å RMS. Testing of an
electroplated gold coating in a simulated space environment indicated that
impingement of any particle of space debris does not extend the damage
regions much beyond the size of the debris or the particle. Further, the gold
coating is relatively hard and easily cleaned.

Now we present the thermal and cryogenic stability data on CVD-SiC. Thermal stability measurements were performed on a 0.25-m polished lightweight SiC substrate. This substrate consisted of a 2.5-mm-thick SiC faceplate with a lightweight SiC back structure deposited by the CVD process. The surface figure and radius of curvature were measured to be 0.1λ rms (root mean square, $\lambda = 0.6328\,\mu m$) and 7.44-m, respectively. This substrate was then heated to 1350°C in a CVD chamber and maintained at that temperature continuously for 60 hours. The substrate was then cooled to room temperature, and its figure and radius of curvature were measured to be 0.41λ rms and 7.48 m, respectively. This change amounts to a figure change of 0.3λ rms and a radius of curvature change of only 0.54%, which is quite small. An optical path difference (OPD) map was also taken before and after thermal cycling and is shown in Fig. 6.13.

The cryogenic stability of the CVD-SiC was assessed on a 2-inch-diameter SiC coupon. The surface figure at room temperature was measured to be 0.045λ rms ($\lambda = 0.6328\,\mu m$). This coupon was cooled to -190°C and then brought back to room temperature, and the surface figure was measured to be $\lambda/125$ rms and $\lambda/70$ rms, respectively. These changes are extremely small and show that CVD-SiC exhibits excellent cryogenic stability.

Next we discuss the optical properties of transparent SiC. Figure 6.14 shows a plot of visible-infrared transmittance of transparent SiC. The sample thickness is about 0.25 mm. The maximum transmittance is about 65%, which is close to the theoretical value for β-SiC. This transmittance corresponds to a specular attenuation coefficient of about 2 and $6.9\,\mathrm{cm^{-1}}$ at wavelength 3 and $0.6328\,\mu m$, respectively. The specular attenuation coefficient at $1.06\,\mu m$ is $1\,\mathrm{cm^{-1}}$. When the contribution of scatter from the attenuation coefficient is taken out, one obtains the absorption coefficient. At wavelength $>3\,\mu m$, the four phonon band is visible [35]. At $3\,\mu m$, the absorption coefficient is about $0.1\,\mathrm{cm^{-1}}$ at 291 K and about $0.4\,\mathrm{cm^{-1}}$ at 912 K. As the wavelength decreases, the absorption coefficient also decreases. The absorption coefficient values at other wavelengths are summarized in Table 6.6.

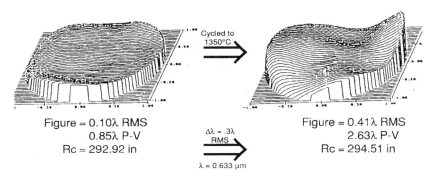

Figure = 0.10λ RMS
0.85λ P-V
Rc = 292.92 in

Cycled to 1350°C

Δλ = .3λ
RMS

λ = 0.633 μm

Figure = 0.41λ RMS
2.63λ P-V
Rc = 294.51 in

Fig. 6.13. Change in figure of 10-inch-diameter lightweight SiC mirror before and after the optic was cycled to 1350°C for 60 hours (from [42]).

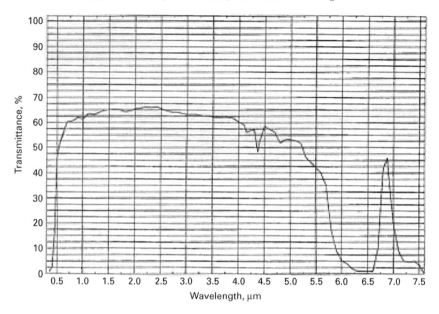

Fig. 6.14. Visible-infrared transmittance of a transparent SiC sample of 0.25-mm thickness.

The emittance of a 1-mm-thick sample of CVD-SiC was computed as a function of wavelength based on the absorption coefficient. The mean emittance at 4–5 μm is about 0.3 at room temperature and 0.5 at 815 K. However, in the 3–4 μm range, the mean emittance is about 0.05 at room temperature and about 0.1 at 815 K. These data show that transparent SiC is a good candidate material for high-heat-flux optics applications, particularly at wavelength less than 4 μm.

The change in refractive index with temperature, dn/dT, for transparent SiC as a function of wavelength was also measured and came out fairly constant in this wavelength range, 2–4 μm and equal to about $37 \times 10^{-6}\,\mathrm{K}^{-1}$.

Table 6.6. Absorption coefficient of transparent SiC.

Wavelength (μm)	Absorption Coefficient (cm^{-1})		
	291 K	609 K	912 K
5.4	14.42	16.84	28.09
4.89	6.0	7.72	16.2
4.47	2.4	3.0	10.11
3.96	1.38	1.53	2.6
3.49	0.52	0.56	0.92
3.22	0.31	0.38	0.65
2.95	0.51	0.17	0.27

6.4 High-Thermal-Conductivity Applications

High thermal conductivity of CVD-SiC along with its many other attractive properties such as high flexural strength, low thermal expansion, and excellent resistance to thermal shock, oxidation, and chemicals makes it an ideal candidate for high-heat-load applications. These applications may arise in several different areas, such as semiconductor processing, optics, electronics, and wear parts [52].

6.4.1 Thermal Management and Semiconductor Processing Applications

High-resistivity CVD-SiC can be used in the electronics industry for thermal-management applications. Because the thermal expansion coefficient of CVD-SiC is compatible with Si, high-power and high-performance Si devices can be produced directly on the SiC substrates. Furthermore, CVD-SiC can be readily metallized with a variety of materials such as silver, gold, TiN, and Mo. This permits fabrication of more complex patterns and structures on SiC substrates.

Even when the resistivity of SiC is not high, it can still be used for thermal-management applications by coating it with a thin layer of CVD diamond to make its surface electrically nonconducting. Both diamond and SiC have low-thermal-expansion coefficients, and thus SiC substrates are compatible with diamond deposition. In addition, diamond has high thermal conductivity, which makes these diamond-coated SiC substrates very effective in thermal management.

In semiconductor processing, CVD-SiC applications include different support components such as susceptors, lifting parts, plates, and flow controls for such processes as RTP (rapid thermal processing), CVD, ion implantation, etching, lithography, and dry, vapor-phase, and wet cleaning. High-heat-load applications include susceptors and rings for processes such as RTP; cantilevers and wafer carriers for oxidation or diffusion furnaces; electrodes, liners, focusing rings, and plasma screens for plasma etch systems; and susceptors for RF heating [53]. Figure 6.15 shows a picture of several CVD-SiC components including RTP rings, susceptors, and gas diffusion plates used in semiconductor processing systems.

Semiconductor processing chambers used to deposit epitaxial films require very high temperatures and extreme cleanliness. The severe temperature excursions, particularly in the rapid thermal processing systems have a tendency to crack or flake the SiC coating on the currently used graphite susceptors. Consequently, bulk CVD-SiC susceptors provide an attractive alternative. In addition, the high stiffness-to-weight ratio of CVD-SiC allows the susceptor to have low weight and good surface flatness. This low mass coupled with low heat capacity and high thermal conductivity keeps the heat ramps rapid and contributes to temperature uniformity over the wafer. The

Fig. 6.15. A variety of CVD-SiC components, including rapid thermal processing rings, susceptors, and gas-diffusion plates, used in semiconductor processing systems.

CVD-SiC susceptors also do not readily degrade during hot HCl cleaning cycles, permit tight susceptor tolerances due to a close thermal expansion match with Si, generate fewer particulates, and can be thermally cycled many more times than other competing materials. The excellent machinability and process reproducibility of CVD-SiC ensures that the parts are fabricated to the same shape with consistent high quality.

In addition to susceptors, slip rings constitute a critical element in RTP reactors. Such a ring surrounds the wafer and is usually slightly offset from its plane. The slip rings serve to make the radial temperature profile more uniform in the wafer. Without a slip ring, the edge of the wafer is hotter than the rest of it during ramping, whereas in steady state, there is a thermal loss at the edge. CVD-SiC provides definite advantages over other materials because of its high thermal conductivity, predictable absorption, and emission characteristics, higher-purity and low-particle generation.

Semiconductor furnaces often employ high temperatures for oxidation or diffusion and use support components such as cantilevers, wafer carriers, support tubes, and paddles. Currently used cantilevers are heavy, conduct heat poorly, and have a coefficient of thermal expansion that is different from silicon. Quartz is relatively fragile, produces more particulates, cannot stand HF in wet processing, and may contain impurities, such as sodium. For all these reasons, CVD-SiC offers an attractive alternative. Currently, efforts are being made to fabricate these components from CVD-SiC.

The resistivity of CVD-SiC can be made low (<0.1 ohm-cm) and tailored in the range of 10–1000 ohm-cm without significantly affecting its other properties. This large range of values makes CVD-SiC very attractive for fabricating a variety of support components in the plasma etch chambers. Gas diffusion plates and focusing rings are made from high-resistivity SiC, whereas the liners and plasma screens are made from low-resistivity SiC. Components made from CVD-SiC are robust and last for a long time in the hostile plasma environment. This reduces equipment downtime and makes this material very competitive in terms of cost of ownership.

Low-resistivity SiC may also be used to make susceptors for coupling RF energy in semiconductor furnaces. The high-thermal-shock resistance of CVD-SiC permits heating these susceptors very rapidly to high temperatures (>1200°C). These susceptors are better than graphite susceptors due to their high purity, extremely low particulate shedding, and low wear rate.

6.4.2 Optics and Wear Applications

In optics, CVD-SiC has been used to fabricate lightweight mirrors, X-ray grazing incidence mirrors, optics standards, and optics baffles [51], [54], [55], [56], [57], [58], [59]. The CVD-SiC mirrors are used in surveillance, high-energy lasers, laser radar systems, synchrotron X-ray equipment and vacuum UV telescopes, large astronomical telescopes, and weather satellites. Active cooling through heat exchanger channels or patterns is employed in optics to manage high heat loads. These patterns can be fabricated on the backside of the CVD-SiC mirror faceplates directly in the CVD chamber by a near-net-shape replication process. Deposition occurs layer by layer on a molecular scale and replicates patterns down to very fine details. This replication of fine features was demonstrated in CVD-SiC during a thermally controlled tertiary mirror (TCTM) program [60]. Figure 6.16 shows a picture of such replication in CVD-SiC. This picture shows a heat exchanger pattern consisting of posts and crosses. The depth of this pattern is about 1 mm, the thickness of crosses is 0.25 mm, the diameter of posts is 0.75 mm, and the spacing between center lines of two adjacent posts is 1.25 mm. This replication was performed on graphite. From Figure 6.16 we see that fine features of the heat exchanger pattern are replicated precisely. There was no evidence of any rounding, pits, holes, or voids on the surface. The cross section of the replicated structure indicated that the sharpness of the replication was maintained throughout the depth of the structure.

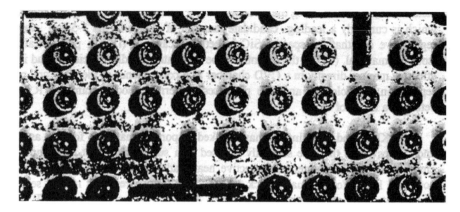

Fig. 6.16. Heat exchanger patterns replicated during CVD-SiC process. Post diameter is 0.75 mm; crosses are 0.25 mm wide, pattern depth is 1 mm, and post spacing is 1.25 mm.

Table 6.7 shows a comparison of important properties of CVD-SiC with other candidate mirror materials. The important figures of merit are: (1) density, ρ; (2) the pressure and bowing distortion parameter E, which is the modulus of elasticity; (3) the thermal distortion parameter κ/α, which is the ratio of thermal conductivity to coefficient of linear expansion; (4) the natural frequency and inertia loading parameter E/ρ; and (5) the thermal stress

Table 6.7. Comparison of important properties of CVD-SiC with other candidate mirror materials.

Material Property	CVD SiC	Mo	Al	Be	ULE 7971	Zerodur
Density, ρ $(\text{kg m}^{-3} \times 10^3)$	3.21	10.2	2.7	1.85	2.20	2.55
Coefficient of thermal expansion $\alpha\,(\text{K}^{-1} \times 10^{-6})$	2.4	5.4	25.0	11.4	0.03	0.15
Thermal conductivity $\kappa\,(\text{Wm}^{-1}\text{K}^{-1})$	325	134	237	216	1.3	6.0
Elastic modulus, E (GPa)	466	250	76	303	67	90
Thermal distortion parameter $\kappa\alpha^{-1}$ $(\text{Wm}^{-1} \times 10^7)$	13.5	2.5	0.95	1.9	4.3	4.0
Inertia loading parameter $E\rho^{-1}\,(\text{Nmkg}^{-1} \times 10^6)$	145	24.5	28.1	164	30.4	35.3
Thermal stress parameter $\kappa\alpha^{-1}\,E^{-1}$ $(\text{WmN}^{-1} \times 10^{-4})$	2.9	1.0	1.25	0.63	6.4	4.4

parameter $\kappa/\alpha E$. It is desirable to have high values for all these figures of merit except density. Thermal conductivity plays an important role in two figures of merit: thermal distortion and thermal stress. Thus, a high value of thermal conductivity increases the value of both these parameters. From Table 6.7 we see that SiC has the highest values of the elastic modulus and the thermal distortion parameter and the second-highest value of the inertia loading parameter. These properties, combined with the fact that SiC is a lightweight material with moderate values of the thermal stress parameter, make it a preferred material for reflective optics applications.

Klein [61] evaluated different candidate mirror materials for cooled high-energy laser applications. He considered a 1-mm-thick faceplate cooled by heat exchanger channels on the back side. The results showed that if laser beam distortions are of concern, the important parameter to consider is the thermal distortion parameter. This study ranked SiC third after diamond and carbon-carbon composites, but ahead of Si, Mo, and Cu. Because diamond is an expensive material that is difficult to scale to large sizes, and carbon-carbon composites have fabrication issues, SiC appears very promising for this application.

The transparent SiC can be used in transmissive optics applications for severe environments associated with high-speed missiles, combustion, space, and laser systems. Klein and Gentilman [62] have ranked important materials for use as windows and domes when they are suddenly exposed to a supersonic flight environment. This environment leads to intense convective heat loads due to rise in temperature of the boundary layer. For the thermally thick case ($B_i > 1$), transparent SiC ranked second after Si_3N_4, but ahead of diamond, sapphire, and AlN. For thermally thin case ($B_i < 1$), the transparent SiC also ranked second, behind diamond, but ahead of AlN, Si_3N_4, and sapphire.

An important application of transparent SiC is its use in laser welding systems and free electron lasers operating at a wavelength of $1.06\,\mu m$. Table 6.8 shows the results of high-power CW Nd:Yag laser irradiation of a sample of transparent SiC for laser-welding applications. The Yag laser was passed through a fiber-optic cable to produce a spot size of $750\,\mu m$. The sample thickness was 0.54 mm. The input power was varied in the range 55–550 watts. We see that even at very high-energy densities there was no appreciable degradation in the transmittance of transparent SiC. Further, after laser irradiation was completed, no visible damage to the transparent SiC sample was observed. In comparison, other competing materials such as sapphire, AlON, CLEARTRAN®, and quartz, did not survive the extreme thermal shock.

In the area of wear, CVD-SiC has been successfully used as a substrate material for making optics molds because of its high value of thermal conductivity, elastic modulus, and flexural strength and its resistance to abrasion, scratching, oxidation, and corrosive materials. The use of CVD-SiC provides a more uniform temperature over the whole surface of glass or plastic

Table 6.8. CW Nd : Yag laser irradiation results: laser wavelength $= 1.06 \,\mu$m; spot size $= 750 \,\mu$m; sample thickness $= 0.54$ mm.

Input Power (W)	On Time (s)	Power Density (KW cm^{-2})	Energy Density (KJ cm^{-2})	Output Power (W)	Transmittance (%)
55	5	12	60	34	62
58	5	13	65	37	64
82.5	5	19	95	52	63
290	5	66	330	177	61
550	3	125	375	—	—

optics, thus minimizing residual stress during lens cool-down. The significant advantages of CVD-SiC are particularly realized when large-area optics molds are used. Further CVD-SiC molds are robust and have been successfully used for fabricating hundreds of optics parts from a single mold.

6.5 Summary and Conclusions

CVD-SiC is a good material for high-heat-flux applications due to its superior thermal, optical, physical, and mechanical properties, particularly its thermal conductivity. The thermal conductivity of CVD-SiC depends on the particular growth method and specific process conditions used for growth. In general, high-thermal-conductivity values are obtained along the columnar grains in a material that has low stacking faults. CVD-SiC has good high-temperature property retention. The flexural strength of CVD-SiC increases slightly and the elastic modulus reduces by about 15% when CVD-SiC is heated to a temperature of 1500°C. The flexural strength of CVD-SiC increased by 37–53% when it was heated to a temperature range of 600°C–1000°C and quenched in water or annealed in flowing N_2 or a vacuum. The electrical resistivity of CVD-SiC was tailored in the range of 0.01–1000 ohm-cm by adjusting the dopant concentration without affecting its thermal conductivity. Due to its superior properties, CVD-SiC is used for thermal management applications in semiconductor and electronic devices and for high-heat-load applications in RTP and plasma etch systems, RF heated susceptors, synchrotron and high-energy-laser mirrors, and optics molds.

References

[1] E. G. Acheson, "Production of Artificial Crystalline Carbonaceous Materials," US Patents # 492,767 (1893); "Composition of Matter for Abrading Articles", US Patent # 572,852 (1896).

[2] W. D. G. Boecker, "Silicon carbide: From Acheson's invention to new industrial products - Dedicated to Prof Dr Hans Hausner on the occasion of his 70th birthday," *Ceram. Forum Int.* **74/5**, 244–51 (1997).

[3] R. F. Davis, "Correlation Among Process Routes, Microstructures and Properties of Chemically Vapor Deposited Silicon Carbide," *Mat. Res. Soc. Symp. Proc.* **168**, 145–57 (1990).

[4] E. Tobin, M. Magida, S. Kishner, and M. Krim, "Design, Fabrication, and Test of a Meter-Class Reaction Bonded SiC Mirror Blank," *SPIE Proc.* **2543**, 12–21 (1995).

[5] C. J. Shih and A. Ezis, "The Application of Hot-Pressed SiC to Large High Precision Optical Structures," *SPIE Proc.* **2543**, 24–37 (1995).

[6] J. R. Block and R. J. Drake, "Silicon Carbide Makes Superior Mirrors," *Laser Foc. World* **25**(8), 97–105 (1989).

[7] V. B. Shield, K. Fekade, and M. G. Spencer, "Near-equilibrium Growth of Thick, High Quality Beta-SiC by Sublimation," *Appl. Phys. Lett.* **62**, 1919 (1993).

[8] M. Shigeta, Y. Fugii, K. Furukawa, A. Suzuki, and S. Nakajima, "Chemical Vapor Deposition of Single-crystal Films of Cubic SiC on Patterned Si Substrates," *Appl. Phys. Lett.* **55**, 1522 (1989).

[9] J. S. Goela and R. L. Taylor, "Rapid Fabrication of Lightweight Ceramic Mirrors Via Chemical Vapor Deposition," *Appl. Phys. Lett.* **54**, 2512–14 (1989).

[10] J. S. Goela and R. L. Taylor, "Fabrication of Lightweight Si/SiC LIDAR Mirrors," *SPIE Proc.* **1062**, 37–49 (1989).

[11] J. S. Goela and R. L. Taylor, "Large-scale Fabrication of Si/SiC LIDAR Mirrors," *SPIE Proc.* **1118**, 14–24 (1989).

[12] M. A. Pickering, R. L. Taylor, J. T. Keeley, and G. A. Graves, "Chemically Vapor Deposited Silicon Carbide (SiC) for Optical Applications," *Nucl. Instrum. Methods*, **A 291**, 95–100 (1990).

[13] J. S. Goela and R. L. Taylor, "Monolithic Material Fabrication via Chemical Vapor Deposition," *J. Mater. Sci.* **23**, 4331–39 (1988).

[14] M. A. Pickering and R. L. Taylor, "Fabrication of Large Mirror Substrates by Chemical Vapor Deposition," *U.S. Air Force Technical Report No.* AFWAL-TR 87–4016 (Wright Patterson Air Force Base, Ohio, April, 1987).

[15] J. Schlichting, "Chemical Vapor Deposition of Silicon Carbide," *Powder Metall. Int.* **12**, 141–47, 196–200 (1980).

[16] K. Nihara, "Mechanical Properties of Chemically Vapor Deposited Nonoxide Ceramics," *Ceram. Bull.* **63**, 1160–63 (1984).

[17] J. R. Weiss and R. J. Diefendorf, "Chemically Vapor Deposited SiC for High Temperature and Structural Applications" in *Silicon Carbide-1973*, ed. R. C. Marshall, J. W. Faust, Jr., and C. E. Ryan. 80–91 (1973).

[18] D. H. Kuo, D. J. Cheng, W. J. Shyy, and M. H. Hon, "The Effect of CH_4 on CVD β-SiC Growth," *J. Electro Chem. Soc.* **137**, 3688–92 (1990).

[19] M. D. Allendorf, C. F. Melius, and T. H. Osterheld, "Thermodynamics and Kinetics of Methyltrichlorosilane Decomposition," *Sandia National Lab Report No.* SAND93-8464, UC-401 (February 1993).

[20] M. D. Allendorf, T. H. Osterheld, and C. F. Melius, "The Decomposition of Methyltrichlorosilane: Studies in a High Temperature Flow Reactor," *Sandia Report No.* SAND94–8524, UC-361 (January 1994).

[21] J. Chin, P. K. Gantzel, and R. G. Hudson, "The Structure of Chemical Vapor Deposited Silicon Carbide," *Thin Solid Films* **40**, 57–72 (1977).

[22] M. G. So and J. S. Chun, "Growth and Structure of Chemical Vapor Deposited Silicon Carbide from Methyltrichlorosilane and Hydrogen in the Temperature Range of 1100 to 1400C", *J. Vac. Sci. Technol.* **A6**, 5–8 (1988).

[23] V. B. Shield, K. Fekade, and M. G. Spencer, "Near-equilibrium Growth of Thick, High Quality Beta-SiC by Sublimation," *Appl. Phys. Lett.* **62**, 1919–21 (1993).

[24] H. S. Kong, J. J. Glass, and R. F. Davis, "Epitaxial Growth of β-SiC Thin Films on 6H α-SiC Substrates via Chemical Vapor Deposition," *Appl. Phys. Lett.* **49**, 1074–76 (1986).

[25] M. Shigeta, Y. Fugii, K. Furukawa, A. Suzuki, and S. Nakajima, "Chemical Vapor Deposition of Single-crystal Films of Cubic SiC on Patterned Si Substrates," *Appl. Phys. Lett.* **55**, 1522–24 (1989).

[26] A. Steckl and J. P. Li, "Epitaxial Growth of β-SiC on Si by RTCVD with C_3H_8 and SiH_4," *IEEE Trans. Electr. Dev.* **39**, 64–74 (1992).

[27] J. S. Goela and R. L. Taylor, "Post Deposition Process for Improving Optical Properties of Chemical-Vapor-Deposited Silicon," *J. Am. Ceram. Soc.* **75**, 2134–38 (1992).

[28] C. H. Chu and M. H. Hon, "Morphology and IR Transmission of β-SiC Synthesized by Chemical Vapor Deposition," *J. Ceram. Soc. Japan* **101**(1), 95–98 (1993).

[29] J. S. Goela, N. E. Brese, M. A. Pickering, and J. E. Graebner, "Chemical Vapor Deposited Materials for High Thermal Conductivity Applications," *MRS Bulletin* **26**, 458–63 (June 2001).

[30] J. S. Goela, L. E. Burns, and R. L. Taylor, "Chemical Vapor Deposition-produced Silicon Carbide having Improved Properties," *US Patent #5,604,151* (Feb. 18, 1997).

[31] J. S. Goela, L. E. Burns, and R. L. Taylor, "Transparent Chemical Vapor Deposited Beta-SiC," *Appl. Phys. Lett.* **64**, 131–33 (1994).

[32] Y. Kim, A. Zangvil, J. S. Goela, and R. L. Taylor, "Microstructure of Transparent SiC from Chemical Vapor Deposition," *Inst. Phys. Conf.* **Ser. No. 137**, Chapter 6, IOP Publishing Ltd., 569–71 (1994).

[33] Y. Kim, A. Zangvil, J. S. Goela, and R. L. Taylor, "Microstructure Comparison of Transparent and Opaque CVD-SiC," *J. Am. Ceram. Soc.* **78**, 1571–79 (1995).

[34] J. S. Goela and R. L. Taylor, "Transparent SiC for Mid-IR Windows and Domes," *SPIE Proc.* **2286**, 46–59 (1994).

[35] S. Anderson and M. E. Thomas, "Infrared Properties of CVD β-SiC," *Proc. SPIE* **3060**, 306 (1997).

[36] J. S. Goela, L. E. Burns, and R. L. Taylor, "Chemical Vapor Deposition-produced Silicon Carbide having Improved Properties," *US Patent #5,612,132* (Mar. 18, 1997).

[37] J. S. Goela and M. A. Pickering, "Optics Applications of Chemical Vapor Deposited SiC," *Proc. SPIE* **CR67**, 71 (1997).

[38] J. S. Goela and M. A. Pickering, "Method for Producing Free-standing Silicon Carbide Articles," *US Patent #6,228,297 B1* (May 8, 2001).

[39] J. Keeley, J. S. Goela, M. Pickering, and R. L. Taylor, "Selective Area Growth in a Vapor Deposition System," *U.S. Patent No. 4,990,374* (February 5, 1991).

[40] J. S. Goela, R. D. Jaworski, and R. L. Taylor, "Method to Prevent Backside Growth on Substrates in a Vapor Deposition System," *U.S. Patent No. 4,963,393* (Oct. 16, 1990).

[41] J. S. Goela and M. A. Pickering, "CVD-SiC Manufacturing Process Reproducibility," *Cer. Eng. Sci. Proc.* **19**, 579–88 (1998).

[42] J. S. Goela, M. A. Pickering, R. L. Taylor, B. W. Murray, and A. Lompado, "Properties of Chemical Vapor Deposited SiC for Optics Applications in Severe Environments," *Appl. Opt.* **30**, 3166 (1991).

[43] M. A. Pickering, R. L. Taylor, J. T. Keeley, and G. A. Graves, "Chemically Vapor Deposited Silicon Carbide (SiC) for Optical Applications," *Nucl. Instrum. Methods*, **A 291**, 95 (1990).

[44] H. Tateyama, H. Noma, Y. Adachi, and M. Komatsu, "Prediction of stacking faults in beta-silicon carbide: X-ray and NMR studies," *Chem. Mater.* **9**, 766 (1997).

[45] V. V. Pujar and J. D. Cawley, "Computer Simulation of Diffraction Effects due to Stacking Faults in β-SiC, I, Simulation Results," *J. Amer. Ceram. Soc.* **80**, 1653 (1997).

[46] V. V. Pujar and J. D. Cawley, "Effect of Stacking Faults on the X-ray Diffraction Profiles of β-SiC Powders," *J. Amer. Ceram. Soc.* **78**, 774 (1995).

[47] A. K. Collins, M. A. Pickering, and R. L. Taylor, "Grain Size Dependence of the Thermal Conductivity of Polycrystalline CVD β-SiC at Low Temperature," *J. Appl. Phys.* **68**, 6510–12 (1990).

[48] C. A. Klein, "Infrared Missile Domes: heat Flux and Thermal Shock," *SPIE Proc.* **1997**, 150–69 (1993).

[49] R. L. Gentilman, "Current and Emerging materials for 3–5 Micron IR Transmission," *SPIE Proc.* **683**, 2–11 (1986).

[50] H. Wang, R. N. Singh, and J. S. Goela, "Effects of Post Deposition Treatments on the Mechanical Properties of a Chemical Vapor Deposited Silicon Carbide," *J. Am. Ceram. Soc.* **78**, 2437–43 (1995).

[51] J. S. Goela, M. A. Pickering, and R. L. Taylor, "Wavelength Dependence of Scatter in Chemical Vapor Deposited SiC," *SPIE Proc.* **1753**, 77–89 (1992).

[52] J. S. Goela, L. E. Burns, and M. A. Pickering, "Chemical vapor deposited SiC for high heat flux applications," *SPIE Proc.* **2855**, 2–13 (1996).

[53] J. S. Goela, N. E. Brese, L. E. Burns, and M. A. Pickering, "CVD-SiC for RTP Chamber Components," IEEE 9th Inter Conf. on Advanced Thermal Processing of Semiconductors, eds. D.P. Dewitt, J. Gelpey, B. Lojek and Z. Nenyei, Sept, 25–29, 2001, p. 217–24.

[54] J. S. Goela, M. A. Pickering, and R. L. Taylor, "Chemical Vapor Deposited β-SiC for Optics Applications," in *Chemical Vapor Deposition of Refractory Metals and Ceramics III*, eds. B.M. Gallois, W. Lee and M.A. Pickering, Material Research Society, (1995) p. 71.

[55] R. L. Gentilman and E. A. Maguire, "Chemical Vapor Deposition of Silicon Carbide For Large Area Mirrors," *Proc. SPIE* **315**, 131 (1981).

[56] J. S. Goela, H. Desai, R. L. Taylor, and S. E. Olson, "Thermal Stability of Chemical Vapor Deposited Beta SiC," *Proc. SPIE* **2543**, 38 (1995).

[57] A. Collins, J. Keeley, M. A. Pickering, and R. L. Taylor, "Investigation of CVD β-SiC Surfaces Produced via a Novel Surface Replication Process," *Mat. Res. Soc. Symp. Proc.* **168**, 193 (1990).

[58] M. A. Pickering and J. S. Goela, "Silicon Carbide Mirror Substrate Replication by Chemical Vapor Deposition," *Rome Laboratory Technical Report No. RL-TR-94-155* (Sept. 1994).

[59] N. Geril, L. Grigley, S. Wilson, and J. S. Goela, "Thin Shell Replication of Grazing Incident (Wolter Type I) SiC Mirrors," *Proc. SPIE* **2478**, 215 (1995).

[60] TCTM Interim Concept Design Review, *Litton Itek Optical Systems* (Feb. 12, 1991).

[61] C. A. Klein, "Materials for High Power Optics: Figure of Merit for Thermally Induced Beam Distortions," *Opt. Eng.* **36**, 1586 (1997).

[62] C. A. Klein and R. L. Gentilman, "Thermal Shock Resistance of Convectively Heated Infrared Windows and Domes," *SPIE Proc.* **3060**, (1997).

7

Chemical-Vapor-Deposited Diamond for High-Heat-Transfer Applications

J.S. Goela and J.E. Graebner

7.1 Introduction

Diamond has always been a material of intense interest for scientists due to its wide range of extreme properties, such as very high thermal conductivity; low expansion coefficient; high hardness, elastic modulus and electrical resistivity; low dielectric constant; high resistance to heat, acids, and radiation; and optical transmission over a wide range of wavelength, from the ultraviolet to the far infrared [1], [2]. These properties make it an ideal candidate material for many applications in the areas of thermal management, optical windows and domes, cutting tools, precious gems, and wear parts. A major obstacle in using natural diamond for many applications is its high cost and availability only in small size and quantity. To overcome these obstacles, scientists have been trying to develop a synthetic route to diamond production that would produce diamond crystals comparable in quality to natural diamond. Initial efforts were focused on developing synthetic diamond by compressing carbon in a high-temperature (1550–2250°C) and high-pressure (50,000–100,000 atmosphere) cell. This technique requires massive equipment, but under suitable conditions it can produce high-quality diamond single crystals.

Interest in diamond became intense in the last twenty years with the discovery that high-quality diamond can be produced by chemical-vapor-deposition (CVD) techniques. CVD diamond exhibits many thermal, optical, mechanical, and electrical properties comparable to natural diamond if the growth conditions are optimized. CVD diamond is a theoretically dense, highly pure, polycrystalline material. The CVD process permits use of near-net-shape techniques to produce components that do not require extensive postdeposition fabrication. Further, the CVD process has been scaled to produce monolithic parts of 0.25-m-diameter and a few mm thick. Large-scale capability reduces CVD diamond cost and makes it a more attractive material for use in different applications. Finally, the CVD process is reproducible. A consequence of reproducibility of CVD diamond is that production and fabrication processes can be developed to yield parts of consistent quality and finish from batch to batch.

Although claims for diamond deposition using low-pressure gases date as far back as 1911 [3], [4], systematic studies of diamond deposition using vapor deposition techniques began in the 1950s in the Soviet Union and United States. Boris Derjaguin and Boris Spitsyn led the early Soviet efforts to deposit diamond on diamond seeds by chemical vapor deposition. These efforts involved thermal decomposition of hydrocarbons and H_2/hydrocarbon gas mixtures at $1000°C$ with no additional activation of gas mixtures. These results were published in a *Scientific American* [5] article but did not generate much enthusiasm in the scientific community because the reported diamond growth rates were quite low (angstroms/hr) and graphitic carbon always codeposited with diamond.

During this early time period, similar research was being conducted in the United States. William Eversole of Union Carbide deposited new layers of carbon atoms (diamond) on the surface of natural diamond seed crystals by decomposition of CO or CH_4 at $900-1100°C$ [6], [7]. This synthesis process, documented in a patent filed in 1958, required many cycles of growth followed by hydrogen etching to remove excessive graphitic deposits [7]. Again the diamond growth rates were very low. In the 1960s and 1970s John Angus and coworkers [8], [9] pursued these diamond-deposition techniques and also obtained diamond deposits on natural diamond powders at low deposition rates.

In the late 1970s and early 1980s, Deryagin's group reported the use of gas activation techniques that eliminated much of the graphite codeposition and resulted in dramatic increases in diamond growth rates [10], [11]. The major Japanese effort also began in the 1970s at the National Institute for Research on Inorganic Materials and many papers were published in the early 1980s describing diamond deposition by hot filament, radio-frequency glow discharge, electron-assisted chemical vapor deposition, and microwave-plasma-assisted processes [12], [13], [14], [15], [16]. These papers generated intense interest in the scientific community, and a large number of studies were initiated with the aim of understanding CVD diamond-deposition processes.

Diamond has a cubic crystallographic structure formed completely from tetrahedrally bonded sp^3 carbon. This should be distinguished from graphite, which is formed completely from trigonally bonded sp^2 hybridized carbons, and lonsdaleite which is a hexagonal form of diamond. The four equivalent sp^3 bonds in diamond and lonsdaleite form strong, uniform, three-dimensional frameworks whereas the graphitic sp^2 bonding creates strongly bonded two-dimensional planes but weak bonding between the planes. In diamond, the three carbons at one end of the bond are staggered with respect to the three carbons at the other end of the bond. In Lonsdaleite, these same carbons eclipse each other, and this causes its structure to be slightly less stable than the diamond structure [3].

Diamond can be distinguished from mixtures of other forms of carbon by measuring the Raman spectrum in the wave number range of $1000-1700 \, cm^{-1}$.

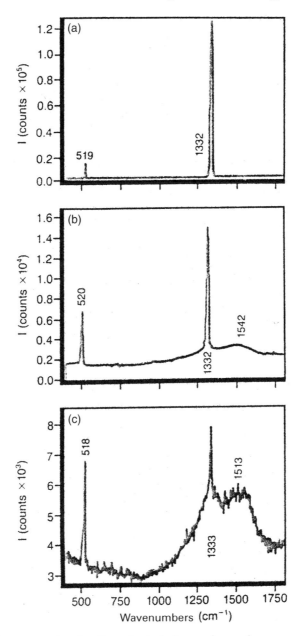

Fig. 7.1. Raman spectra for three typical diamond samples vapor deposited on single crystal silicon (a) highly perfect diamond film, (b) film of intermediate perfection, and (c) film containing appreciable amounts of sp^2 carbon. Note the intensity scale change in each of the three spectra. The $519\,\mathrm{cm}^{-1}$ line is from the silicon substrate, the $1332\,\mathrm{cm}^{-1}$ line is characteristic of diamond, the broad peak in the 1500–$1600\,\mathrm{cm}^{-1}$ range is characteristic of the sp^2-type disordered carbon. (Reprinted with permission of the American Ceramic Society, www.ceramics.org. © 1989 [3].)

Highly perfect diamond shows a very sharp peak at or around $1332\,\mathrm{cm}^{-1}$. The disordered sp^2-type carbons show broad peaks in the 1500–1600 wave number range. Diamond-like carbon films that have very fine, nanosize crystallites would not show the 1332 wave number peak. This technique is very sensitive for identifying pure diamond sp^3 bond because the ratio of Raman scattering efficiency for sp^2 versus sp^3 carbon is 50 [3]. Figure 7.1 shows three Raman spectra for different quality diamond films deposited on silicon substrates. The $519\,\mathrm{cm}^{-1}$ line is from the silicon substrate.

In this chapter, we present a brief review of the CVD diamond process and important properties relevant for high-thermal-conductivity applications. In Sect. 7.2, the CVD processes used to produce high-thermal-conductivity diamond are presented. Many applications may require not only high thermal conductivity but also optimum values of other properties such as flexural strength, optical transmission, electrical resistivity, and chemical purity. Consequently, important physical, thermal, mechanical, optical, and electrical properties of CVD diamond are provided in Sect. 7.3. The high-thermal-conductivity applications of CVD diamond are discussed in Sect. 7.4. Finally, summary and conclusions are presented in Sect. 7.5.

7.2 Diamond Synthesis by CVD

The chemical-vapor-deposition process involves heterogeneous reaction of a gas-phase compound or compounds on or near a substrate surface to produce a solid deposit. The CVD technique for producing diamond involves activating a mixture of H_2 or O_2 and carbon-containing gases to produce diamond on a heated solid surface in a deposition chamber (Fig. 7.2). The carbon-containing gases could be hydrocarbons, such as CH_4, C_2H_2, or CO, CO_2; various alcohols; or acetone. The activating source could be a hot filament, a plasma (DC, RF, or microwave), a combustion flame (oxyacetylene or plasma torches), an optical pumping source, or a laser. The last two techniques are more recent and offer the advantages of lowering the diamond-growth temperatures and driving diamond growth through selective reaction pathways. Although most diamond-deposition processes use excess H_2 in the reaction mixture, CVD diamond has also been produced by processes that contain large amounts of O_2. In the CVD process, the diamond deposition occurs at those temperature and pressure conditions under which graphite is the stable form of carbon. This occurs because activation of the reactant gas mixtures drives a complex chemistry that inhibits graphite formation and promotes diamond growth.

Two of the popular CVD processes that also produce high-quality diamond deposits are the microwave plasma-assisted CVD process (MPACVD) and the hot-filament method (HFCVD). The MPACVD technique was first proposed

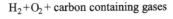

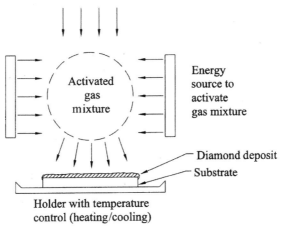

Fig. 7.2. A schematic of a generic CVD process to deposit diamond.

by Kamo et al. [13]. A schematic diagram of a typical tubular microwave reactor is shown in Fig. 7.3. It consists of a magnetron source that produces microwaves at a frequency of 2.45 GHz. These microwaves are monitored by a power monitor, tuned with a stub tuner, and then pass through a metallic waveguide. A plunger is attached at the end of the waveguide to minimize the reflected power. A quartz tube passing through a sleeve and a hole in the waveguide serves as the diamond deposition chamber. The end of the quartz tube is connected to a vacuum pump to maintain low pressure in the deposition chamber. The substrate is mounted in a holder and placed at the intersection of the quartz tube and the waveguide. The substrate temperature is controlled by heating the holder either with the microwaves or with an independent resistive heater. The substrate temperature is measured with either a thermocouple or a pyrometer. When a mixture of H_2 and hydrocarbon (such as CH_4) is passed through the quartz tube, microwaves create a plasma in the quartz tube and activate the gas mixture causing diamond to deposit on the substrate.

Typical deposition conditions for MPACVD are as follows: tube diameter <4.5 cm, 0.2–4 vol% CH_4 or other hydrocarbon diluted in H_2, chamber pressure = 1–400 Torr, substrate temperature = 500–1000°C, microwave power = 100 W–1.5 KW and substrate sizes 2–3 cm in diameter. The typical growth rates are 0.1–30 μm/hr [17]. Since the first use of a microwave reactor, great advancement has been made in the development of various types of microwave plasma reactors aimed at improving the quality and size of the diamond deposition. Some of the microwave machines used include bell jar reactors, plasma jet, and disk reactors, and ellipsoid, surface-wave, and magneto microwave plasma reactors [18]. Using advanced microwave plasma machines, the parameters for diamond deposition have been extended over a

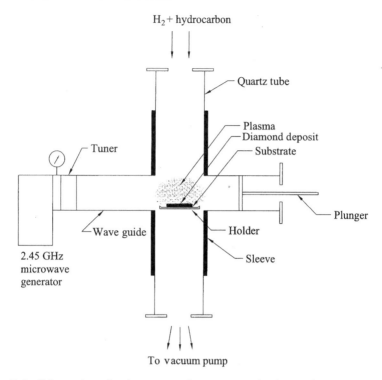

Fig. 7.3. Schematic of microwave plasma-assisted chemical-vapor-deposition-system (from [13]).

wide range as follows: substrate size ≤ 30 cm, growth rates ≤ 1 g/hr, chamber pressure $= 0.01$–760 Torr, and microwave power ≤ 60 KW.

A hot-filament system is popular because it is a relatively simple and inexpensive system (Fig. 7.4). It consists of a small vacuum chamber, 55–80 mm diameter, which could be made from quartz, pyrex, or alumina tubes. On one end of this tube is mounted a filament of suitable material such as tungsten, tantalum, rhenium, or platinum. The filament is heated to a temperature of 1800–2300°C. A mixture of 1% CH_4 in H_2 is passed on this filament to preheat and dissociate the gas mixture. The substrate of a suitable material such as Si or Mo is mounted in a holder and is placed 3–10 mm away from the filament. The substrate temperature is maintained in the range of 600–1000°C with a resistance heater. The substrate and filament temperatures are measured with a thermocouple and optical pyrometer, respectively. The chamber pressure is maintained in the range of a few to hundreds of Torr. The gas flow rate is in the range of 0.1–1 standard liters per minute. The deposition time depends on the thickness of the diamond film required and typically ranges from one to a few hours. Typical growth rates are 1–5 μm per hour.

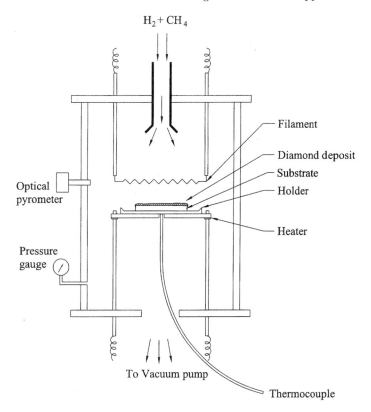

$H_2 + CH_4$

Filament

Diamond deposit

Substrate

Optical
pyrometer

Holder

Heater

Pressure
gauge

To Vacuum pump

Thermocouple

Fig. 7.4. Schematic of a hot-filament-assisted chemical-vapor-deposition diamond growth system.

Important issues with the hot-filament process are scaling to large sizes, low growth rates, diamond film contamination with hydrogen atoms and filament material, and diamond film quality, uniformity, and near net shaping. Over the past twenty years considerable technological developments have taken place that have provided innovative solutions to these issues. With current technology hot-filament systems can provide good-quality diamond films on substrates up to 12 inch diameter and growth rates up to $19\,\mu m/hr$ for a single filament system and $5\,\mu m/hr$ for multifilament systems and some near net shaping on dome shapes [2], [19].

Hydrogen plays a critical role in the deposition of diamond at temperatures at which graphite growth is favored. Hydrogen atoms are produced when the gas mixture is "energized" either thermally or with electron impingement. The hydrogen atoms promote the diamond deposition process as follows: (1) H atoms react with hydrocarbon molecules to create reactive radicals such as CH_3, which react on the substrate surface to produce C–C bonds necessary for diamond growth, (2) H atoms terminate the "dangling" carbon bonds on the growing diamond surface and thus prevent them from forming a graphite-like surface, and (3) atomic hydrogen etches away any graphite that is codeposited with diamond.

The substrate material and its surface preparation are critical for growing continuous, good-quality diamond films. Single-crystal silicon has been the most popular substrate material for growing CVD diamond films because it has a high melting point ($1410°C$) and a low thermal-expansion coefficient, which is more closely matched to diamond than most other substrate materials. It is also relatively inexpensive and readily available. Furthermore, after diamond deposition, silicon can be removed by either polishing or chemical etching to form free-standing diamond. Diamond deposition on Si involves formation of a thin and localized carbide interfacial layer on which diamond then grows. On some substrates, such as Fe, Ni, and Ti, carbon is very reactive and forms thick carbide interfacial layers [1]. These thick carbide layers can affect the mechanical properties and thus the quality of the diamond deposit. Tungsten and Mo are two other substrate materials that perform similarly to Si and can yield good-quality diamond films. Other substrate materials used for diamond deposition are SiC, WC, Cu, diamond, and SiO_2.

In diamond deposition, nucleation usually occurs on surface defect sites such as scratches, digs, protrusions, or steps. These sites are generally created by polishing the substrate surface mechanically or by ultrasonic agitation using the diamond powder. Other methods of creating defect sites, such as chemical etching or polishing with alumina powder, do not provide as high nucleation density as polishing with diamond powders. On untreated surfaces, the nucleation density is usually very low and the crystal size is relatively large. Over the years diamond nucleation has been enhanced on nondiamond substrates by using different techniques such as abrading the substrate surface with powders of diamond and metals (W, Ta, Fe, Ti, Mo), using different forms of carbon (clusters, fullerenes, fibers, graphitic, amorphous), applying a negative bias to the substrate, coating silicon with a catalytic material such as iron, and electrostatic seeding of diamond nanoparticles [20].

7.2.1 Postdeposition Processing

The CVD deposited diamond films usually have nonuniform thickness, are quite rough on the growth surface, have low thermal conductivity on the substrate surface, and are attached to the nondiamond substrate. Consequently, postdeposition processing which may include substrate removal, cutting, grinding, lapping, and polishing, is required to make it suitable for thermal management and other applications. Mechanical polishing and chemical etching processes are used to remove the silicon substrate. Because of the extreme hardness of diamond, cutting of diamond with a diamond saw is a very slow process. A better process of diamond cutting is the use of high-power lasers such as an Nd-YAG laser in combination with a reactive gas such as O_2. The laser beam can be focused to a very fine spot and results in heating and oxidizing the diamond as it is moved along the cut. The carbon in diamond reacts with oxygen to form CO_2 and CO leaving behind a clean cut. High-power pulsed lasers such as an Nd-YAG laser operating at 1.06 micron and 0.532 micron have also been used to drill small holes of diameter 100–150 micron in diamond [20].

To make the diamond surface smooth it is often required to remove 20 to 50 microns or more material from its two surfaces. Many techniques for lapping and polishing of diamond have been developed over the years [21]. Because materials harder than diamond are not available, diamond is usually machined and polished using diamond powder, but this is a very slow process and is not particularly suitable for removing large amounts of material. Oxidizing chemicals such as KOH have been used to enhance the removal rate in the mechanical polishing process. Hot metal lapping on an iron wheel at 900°C in a hydrogen atmosphere has been used successfully because of the high solubility of carbon in iron. Removal of carbon by an oxygen plasma has the disadvantage of preferential etching at grain boundaries. Laser ablation and ion beam polishing are noncontact techniques that have been used for smoothing and patterning of both flat and curved surfaces but their equipment costs are quite high.

Another technique that appears to be well suited to batch processing with high rates of removal makes use of the diffusion of carbon into certain metals, either solid or molten as shown by Jin and coworkers. Free-standing diamond films, for example, can be sandwiched between foils of iron and heat treated under weight at 900°C for 48 hours in an argon atmosphere, resulting in a 100-µm reduction in thickness (1 µm/hr) [22]. Manganese powder can alternatively be used [23] because of its higher solubility for carbon [24]. Some mechanical fine polishing is needed after the thinning process to obtain flat surfaces for thermal applications. Molten rare earth metals such as cerium or lanthanum provide not only higher solubility of carbon, but also intimate contact because of the excellent wettability between diamond and the molten metal [25], [26]. Low-cost mischmetals (a mixture of rare earth metals) [27] or rare earth/transition metal alloys with eutectic melting temperatures [28] have also been used. A very fast diamond-etch rate in excess of 50 µm per minute in thickness reduction has been achieved using molten mischmetals.

Another technology that is important for using diamond for high-thermal-conductivity applications is bonding. For heat-spreader applications, diamond must be connected very well thermally to the heat-producing device and the ultimate heat sink. Several techniques such as applying epoxy adhesive, soldering, or brazing have been proposed for bonding of diamond. The use of epoxy adhesive provides only moderate thermal contact, even if the epoxy is filled with metal powder. Low-temperature solders provide better thermal contact by a factor of ten, but they do not adhere well to diamond and have low thermal conductivity. Strong carbide-forming elements such as Ti, Zr, V, Nb,

Ta, Cr, and Si generally provide reliable metal-diamond bonding. For example, a metallization scheme for solder bonding of laser diodes [29] to diamond uses a three-layer configuration of Ti/Pt/Au. The Ti layer provides bonding to diamond, the Pt layer provides a diffusion barrier, and the gold overcoat serves as a protective layer to minimize platinum-solder reactions and also as a bond layer to a Au-Sn eutectic solder. The thickness of sputter-deposited metallization layers on diamond are, for example, 100 nm Ti, 200 nm Pt, 500 nm Au, and 2.5 µm of Au-20Sn eutectic solder. The solder bonding of the laser diode onto the diamond heat spreader is carried out by rapid thermal annealing (e.g., at 300–350°C for several seconds).

Brazing with metals is another technique that can have higher thermal conductivity than low-temperature solders but the processing is carried out at higher temperatures [21]. Even better thermal contact can be made if the device is grown on one side of a Si wafer and the diamond is deposited on the other side with careful choice of nucleation and growth conditions [30], [31]. This configuration places the source of heat in very close contact with high-thermal-conductivity diamond to achieve maximum cooling efficiency.

7.3 Properties of CVD Diamond

CVD diamond has been extensively characterized for important physical, mechanical, optical, thermal, and electrical properties. Table 7.1 lists important properties of CVD diamond [32], [33], [34], [35], [36]. In general, high-quality CVD diamond samples have exhibited many properties such as transmission, refractive index, thermal expansion coefficient, hardness, and thermal conductivity close to those of natural Type IIa diamond, which is the purest natural diamond. Diamond has a very wide indirect band gap of 5.5 eV, which makes it optically transparent over a wide wavelength range from 0.225 µm in the UV to far infrared beyond 20 µm. However there are 2-phonon and 3-phonon absorption bands in the wave number range of 1332–3996 cm^{-1}, which makes it unsuitable for window applications in the shortwave infrared region (3–5 µm). The elastic modulus and Poisson ratio for randomly oriented polycrystalline diamond can be calculated by appropriately averaging the single-crystal diamond data over all orientations. This yielded a modulus of 1140 GPa and a Poisson ratio of 0.069, which are consistent with the measurement of biaxial modulus, E/(1-Poisson ratio), of randomly oriented CVD diamond deposits [32], [33].

The CVD diamond is a brittle polycrystalline material, and therefore it is susceptible to flaw-induced fracture. The strength of CVD diamond from three different suppliers was measured with a ring-on-ring fixture and came out to be in the range of 200–400 MPa [34]. This strength is less by an order of magnitude than the tensile strength for natural diamond (∼3 Gpa) [35] and is attributed to the presence of microcracks, residual stresses, and

Table 7.1. Important properties of CVD diamond at room temperature

Property	Average Value
Density (g cm^{-3})	3.51
Hardness (Kg mm^{-2})	9000
Fracture toughness (MPa.m$^{0.5}$)	5.3–8
Band gap	5.5 eV
Elastic modulus, GPa	1140
Flexural strength, MPa	>200
Poisson ratio	0.069
Thermal expansion coefficient (10^{-6} K^{-1})	0.8
Thermal conductivity (Wm^{-1}K^{-1})	≤2300
Heat capacity (Jkg^{-1}K^{-1})	640
Electrical resistivity (ohm-cm)	>1.0×10^{13}
Dielectric constant (35–50 GHz)	5.7
Dielectric strength (V m^{-1})	1.0×10^{7}
Loss tangent (35 GHz)	<0.00015
Refractive index @ 10 μm	2.38
Optical absorption coefficient	
@ 8–12 μm	0.1–0.3
Emissivity	
@ 8–12 μm, 300–500°C	0.02–0.03

other flaws in CVD diamond. In addition, no degradation in flexural strength was measured when the temperature increased from 20°C to 1000°C. Klein [33] also analyzed the strength of CVD diamond based on the Weibull distribution and reported similar low values—an average flexural strength of 398 MPa and 160 MPa when the nucleation or the growth surface is in tension, respectively. In addition, the spread in strength data was quite large, which yielded a Weibull modulus, m, of 2.6 and 4.70 when nucleation and growth surfaces were in tension, respectively. Savage et al. [36] measured CVD diamond fracture stress by 3-point bend testing and concluded that the fracture stress decreases as the thickness of the diamond sample decreases (≤2.5 mm) and grain size increases (16–400 μm). In addition, larger values for Weibull modulus ($m = 11$) when nucleation side is in tension were obtained indicating more consistent fracture stress values for CVD diamond.

7.3.1 Thermal Conductivity of Diamond

At temperatures above approximately 50 K, diamond has the highest thermal conductivity of any known material with the possible exception of carbon nanotubes [37]. At room temperature, gem-quality diamond exhibits values of thermal conductivity κ in the range of 2000–2500 W/mK, which is 5 to

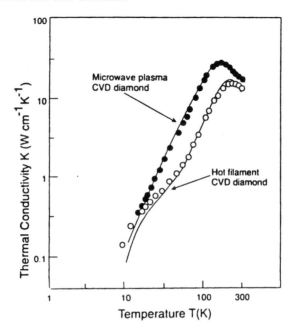

Fig. 7.5. Thermal conductivity of microwave-assisted and hot-filament CVD diamond samples (from [40] with permission).

6 times higher than that of copper [38], [39]. Lower-optical-quality diamond has correspondingly lower thermal conductivity. The particular conditions of growth determine the quality of the diamond, which can vary from black (with $\kappa \sim 300$ W/mK) to clear (with values greater than 2000 W/mK). The quality is generally highest with the lowest growth rate, but it depends on many other parameters as well, such as substrate temperature, gas pressure and composition, and growth method. Morelli et al. [40] measured and compared the thermal conductivity of diamond produced from microwave plasma and hot-filament techniques. Figure 7.5 shows the thermal conductivity data fitted to a Debye model. In the temperature range of 10–300 K, the microwave plasma diamond sample followed the Debye model and showed thermal conductivity higher than the hot-filament diamond sample. The latter sample also showed slight deviation from the Debye model in the temperature range of 20–60 K, and this was attributed to phonon scattering from defects in the hot-filament diamond sample that are not present in the microwave plasma diamond sample.

The thermal conductivity of diamond depends on the grain size [34], [41]. At low temperatures the mean free path for phonons in polycrystalline diamond is 4 to 10 times greater than the grain size [42]. Above 500 K, the phonon mean free path is smaller than the grain size and the thermal conductivity does not vary much with temperature. This suggests that above 500 K, good-quality CVD diamond should have the same thermal conductivity as Type IIa diamond. Reference [34] recommends the following equation for estimating the thermal conductivity of CVD diamond or Type IIa diamond in the range of 300–1200 K:

$$\kappa = (2.833 \times 10^6)/T^{1.245}, \tag{7.1}$$

where κ has the units $Wm^{-1}K^{-1}$ and T is in Kelvin. Equation (7.1) yields $\kappa = 2340\,Wm^{-1}K^{-1}$ at 300 K and $416\,Wm^{-1}K^{-1}$ at 1200 K.

7.3.1.1 Local Thermal Conductivity. The thermal properties of CVD diamond are strongly dependent on the microstructure of the material [43]. Diamond growth begins by the nucleation of individual crystallites at random spots on the substrate, followed by competitive growth as the crystallites enlarge and merge. Those crystallites that happen to be oriented with their fastest crystallographic growth direction normal to the plane of the substrate eventually dominate, so that a strong columnar texture develops with the long axis of the columnar grains oriented normal to the film (Fig. 7.6). The average in-plane dimensions of the grains increase more or less linearly with the distance z from the substrate. With such a microstructure, the thermal conductivity κ_{par} for heat flowing parallel to the film is different from the conductivity $\kappa_\perp$ for heat flowing perpendicular to the film, and the thermal conductivity is thickness-dependent, as described below.

As most methods for measuring thermal conductivity provide a value that is averaged over the full thickness of the sample, special techniques are required to determine the z-dependence of the local thermal conductivity [44]. Slicing the CVD diamond into thin layers at various values of z is a possible approach but is very difficult because of the hardness of diamond and the potential damage during such processing. Instead, five samples of thickness 14, 48.5, 90.5, 145, and 285 μm were prepared by microwave plasma CVD, all under the same conditions and differing only in thickness. The in-plane thermal conductivity κ_{par} of each sample was measured with a DC heated-bar technique with an accuracy of 1–2%. The observed thermal conductivity κ_{par}^{obs} is an average over the thickness of a sample and increases with sample thickness Z as shown in Fig. 7.7 (inset). A local conductivity was deduced using the expression $\kappa_{par}^{local}(z) = \partial\left(Z\kappa_{par}^{obs}\right)/\partial Z$.

The perpendicular thermal diffusivity was measured with a fast version of the laser flash technique [45], [46] for four samples, $0.5 \times 1\,cm^2$ in area, with average thicknesses of 28.4, 69.1, 185, and 408 μm. The diffusivity values were then converted into thermal conductivity by multiplying with the heat capacity per unit volume, ρC. The deduced thermal conductivity $\kappa_\perp^{obs}$ is an average over sample thickness Z and increases with it even more rapidly than κ_{par}^{obs}. The deduced value of thermal conductivity was converted to local values $\kappa_\perp^{local}$ as follows: Because the heat pulse encounters successive layers in series,

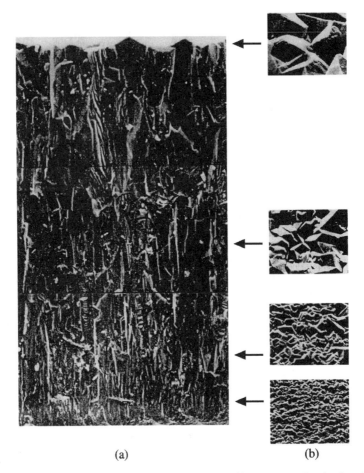

(a)   (b)

Fig. 7.6. (a) Cross-sectional SEM micrograph (fracture surface) showing the dependence of grain size on the height above the substrate, (b) micrograph of the top surface of four specimens of different thickness, illustrating the growth of the grain size with thickness (from [43] with permission).

one can define the resistance per square $R = Z/\kappa_\perp$ for a sample thickness Z. For two samples of thicknesses Z_i and Z_k ($Z_k > Z_i$), the extra resistance per square of sample k compared with sample i is $Z_k/\kappa_{\perp,k} - Z_i/\kappa_{\perp,i}$, which can be defined as the average local resistance per square at the average height $(Z_i + Z_k)/2$. The local conductivity at the average height is then given by

$$\kappa_\perp^{\text{local}} = \frac{(Z_k - Z_i)}{\left(\dfrac{Z_k}{\kappa_{\perp,k}} - \dfrac{Z_i}{\kappa_{\perp,i}}\right)}.\tag{7.2}$$

The conductivities for both heat flow directions are shown in Fig. 7.8. Both local conductivities extrapolate to $2300 \pm 200\,\mathrm{W\,m^{-1}K^{-1}}$ for large z ($> 350\,\mu\mathrm{m}$)

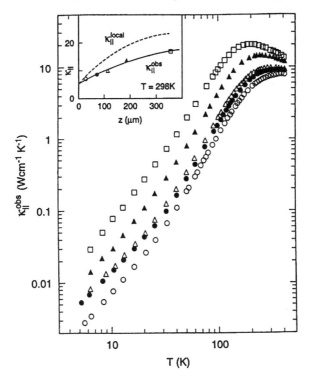

Fig. 7.7. κ_{par} (T) for the five samples of different thicknesses. The inset shows the derivation of the local conductivity at 298K (from [45]).

and $550\,\mathrm{Wm^{-1}K^{-1}}$ for small z ($\sim 5\,\mu\mathrm{m}$). A strong dependence on z for both components of conductivity is evident in the figure (i.e., a factor of at least four comparing the conductivity at large z with that at small z ($\sim 20\,\mu\mathrm{m}$)). An anisotropy $\kappa_{\perp}^{\mathrm{local}}/\kappa_{\mathrm{par}}^{\mathrm{local}} \sim 2$ for z in the range 30–100 $\mu\mathrm{m}$ is also observed. Intuitively, it seems reasonable that the fine-grained less-perfect material near the substrate surface should have lower thermal conductivity than the large-grained more-perfect material for $z \sim 300\,\mu\mathrm{m}$, and heat should flow more easily in the direction of the columns than across them.

A more detailed analysis reveals information about the type and location of the defects responsible for thermal resistance. The in-plane conductivity κ_{par} of the samples was measured [45] from above room temperature to liquid helium temperature and analyzed to deduce a local conductivity as a function of temperature (Fig. 7.9). The phonon-scattering model of thermal conductivity was then applied to the data in Fig. 7.9 on the assumption [46] that the thermal conductivity was limited by a number of distinct phonon-scattering mechanisms: phonon-phonon (intrinsic) scattering, atomic point defect scattering, extended defect scattering, dislocation scattering, and

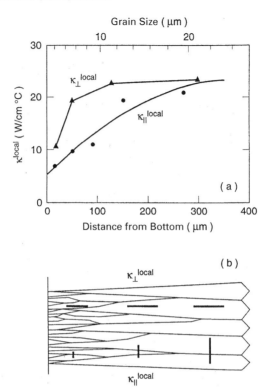

Fig. 7.8. (a) Local thermal conductivity versus height (distance from the bottom) for heat flowing parallel ($\kappa_{||}$) or perpendicular ($k_\perp$) to the plane of the specimen, (b) schematic illustration of the nonuniform anisotropic microstructure in CVD diamond films (from [43]).

clean-grain-boundary scattering. The results are presented in Fig. 7.10 as approximate partial resistivities due to each of these mechanisms.

The general trend toward lower extrinsic thermal resistance with increasing z is consistent with the increasing conductivity with z in Fig. 7.8. The point defect scattering is partly due to the 1% ^{13}C isotopic impurity that occurs naturally in the otherwise ^{12}C carbon. The rest of the point defect scattering is probably not due to the likely suspects, hydrogen and nitrogen, on the basis of infrared absorption and other studies on the same samples, but is more likely due to vacancies [47], [48], [49], [50]. The extended defects are roughly 1.4 nm in diameter, as determined by the broad dip in the $\kappa_{par}(T)$ data in Fig. 7.9 and may be the high-order twin intersections that often occur in diamond. The scattering at clean grain boundaries for the grain size observed is not strong enough to account for a large portion of the total resistivity.

Quantitative comparison of the various contributions to the resistivity in Fig. 7.10 shows that the factor-of-two anisotropy for z in the range ~30–100 μm is difficult to explain unless the (non-^{13}C) point defects and,

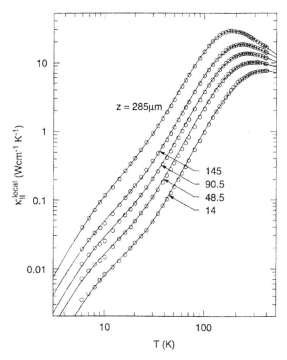

Fig. 7.9. Local thermal conductivity versus absolute temperature at five different heights z above the substrate. The curves are fits of a phonon-scattering model of thermal conductivity with the strengths of various scattering mechanisms adjusted to fit the data (from [45]).

perhaps, the dislocations are located preferentially at or near grain boundaries. Then the conductivity along the grains can be substantially higher than the conductivity across the grains. This dirty-grain-boundary model has been expressed mathematically by Goodson and coworkers with the assumptions that all extrinsic defects are located at grain boundaries and the number density of defects per unit area of the grain boundary is independent of z [51], [52]. With all extrinsic scattering mechanisms parameterized by the grain size, d, the standard phonon-scattering calculations for room temperature yield the solid line in Fig. 7.11. Also shown are the corresponding plot for clean grain boundaries and data from nine separate reports [53]. The clean- and dirty-grain-boundary models both approach $2500\,\mathrm{W m^{-1} K^{-1}}$ at large d, where boundary scattering is negligible compared with intrinsic scattering, but show a large difference as the grain size becomes increasingly smaller. For both plots, the effect of grain size does not reduce the conductivity significantly until the grain size becomes smaller than $\sim 10\,\mu\mathrm{m}$. Because the mean free path in perfect diamond at room temperature is of order $0.3\,\mu\mathrm{m}$, grain size limitation to thermal conductivity occurs when the grain is on the order of a few micrometers or less. Although the experimental data show significant scatter, the

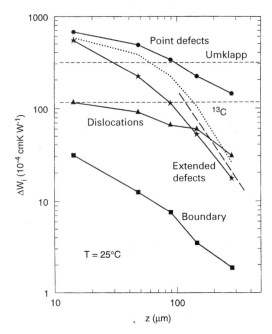

Fig. 7.10. Approximate partial thermal resistivities deduced from the fits to the data of Fig. 7.9. The thermal resistance due to each of the four extrinsic mechanisms decreases monotonically with height z above the substrate. Intrinsic scattering mechanisms (scattering of phonons) from other phonons [umklapp] or from ^{13}C are independent of z (from [45]).

model gives a reasonably good fit over a wide range of grain sizes. Deviations can be understood as cleaner or dirtier grain boundaries, depending on the growth conditions.

7.3.1.2 Thermal Conduction Near Diamond-Substrate Interface. For many applications in thermal management, the thermal resistance at or near the interface between CVD diamond and the substrate is of central importance because it can become a serious bottleneck for heat flow into or out of the diamond. Anisotropy of the in-plane and normal thermal conductivity can become even more severe for diamond layers that are very close to the substrate. Several researchers [31], [54], [55] measured the thermal resistance of thin ($<2.7\,\mu$m thick) diamond films deposited on silicon. Measurements close to the interface required special techniques to determine the thermal resistance. These techniques included: (1) the DC step heating of patterned metal microbridges on the diamond surface and simultaneous temperature measurement by monitoring the electrical resistance of the heater; (2) a laser heating method in which a metal film on the diamond absorbs a short (6 ns) wide-area laser pulse and its temperature is monitored for several microseconds by means of the reflection of a second , continuous–wave laser; and (3) a transient thermal grating laser-heating technique that measures thermal diffusivity

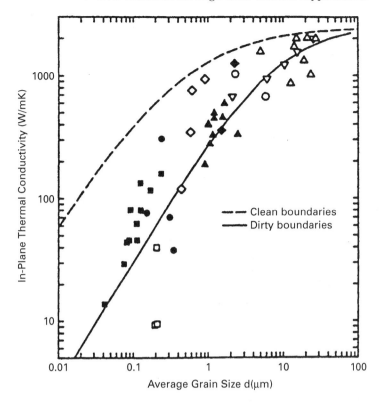

Fig. 7.11. In-plane thermal conductivity $\kappa_\parallel$ at room temperature versus average grain size for CVD diamond from nine reports. The dashed line is the thermal conductivity expected with clean grain boundaries and the solid line is calculated by Goodson et al. [51] using the dirty-grain-boundary model, which assumes that all defects are located at grain boundaries (from [53]).

parallel and perpendicular to the plane of the film. Goodson et al. [54] measured the normal thermal resistance $R_\perp$ of diamond films of three different thicknesses, 0.2, 0.5, and 2.6 μm, and then calculated the effective thermal conductivity $\kappa_\perp$ by dividing film thickness by the thermal resistance. The effective conductivity increased with thickness from 12 to 22 to $74\,\mathrm{W m^{-1}\,K^{-1}}$. These values are much smaller than the values obtained earlier for diamond films of thickness 14 μm and higher and could be due to the presence of a highly imperfect region near the silicon-diamond boundary that exhibits large local resistance. Verhoeven et al. [31] found that the effective boundary resistance at the diamond-silicon interface depends on the grain size at the interface. This resistance can be improved significantly, in some cases by an order of magnitude, by selecting process conditions that yield low nucleation density and large grains at the interface. For instance, the average effective boundary resistance decreased from $2.4 \times 10^{-7}\,\mathrm{m^2\,K\,W^{-1}}$ to $1.2 \times 10^{-8}\,\mathrm{m^2\,K\,W^{-1}}$ when the grain size increased by an order of magnitude from 10 nm. Excessive reduction in nucleation density, however, can also result in voids that reduce thermal conduction.

In another study, Verhoeven et al. [55] used the laser heating and transient thermal grating techniques to measure thermal diffusivity both parallel and perpendicular to the plane of the diamond films grown on Si. Figure 7.12 shows the expected nearly isotropic conductivity for the random nanometer-size grains ($\kappa_\perp/\kappa_{par} = 0.8$). Highly oriented samples exhibited very high anisotropy (a factor of 10–20), especially for those samples with a thin layer of β-SiC separating the diamond and the silicon, and a decrease in boundary resistance by two orders of magnitude. Such isotropy suggests that the dominant scattering of phonons still takes place at dirty grain boundaries or at anisotropic scattering centers, even in the first few microns of these high-quality thin films.

7.3.1.3 Thermal Conductivity of Isotopically Enriched Diamond. Thermal conductivity in isotopically enriched synthetic single crystals of diamond

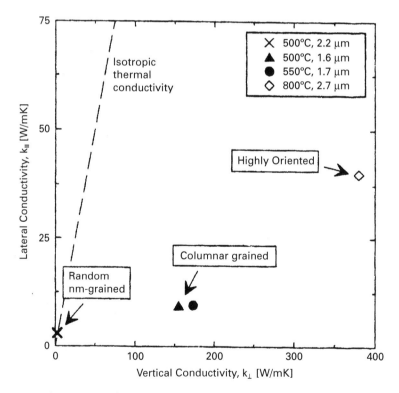

Fig. 7.12. Anisotropy of the thermal conductivity in thin films of CVD diamond on silicon. The dashed line indicates isotropy, $\kappa_{||} = \kappa_\perp$, and emphasizes the very large anisotropy observed for the columnar-grained and the highly oriented (quasi-epitaxial) films. From [55].

have shown [56], [57], [58], [59] a dramatic increase ($\sim$30%) in the room-temperature thermal conductivity as the ^{13}C concentration is reduced from the natural abundance of 1.1% to 0.07%. This has been attributed [58], [59], [60], [61], [62], [63] to the importance of normal phonon-phonon scattering processes, which tend to enhance the effect of Rayleigh scattering from the point defects present (^{13}C atoms). This unusually high sensitivity of the thermal conductivity to point-defect isotope scattering is usually observable only in high-purity single crystals [64]. It allows high-quality diamond, already a remarkable conductor at its typical room-temperature value of 22–25 Wcm^{-1}K^{-1} to be improved to 30–33 Wcm^{-1}K^{-1} by isotopic purification.

The polycrystalline diamond made by the CVD process has also exhibited significant improvement in thermal conductivity due to isotopic enrichment from ^{13}C to ^{12}C. Isotopically enriched (0.055% ^{13}C) diamond plates (650 μm thick) were prepared by microwave-enhanced CVD using ^{12}C-enriched methane and both the in-plane and perpendicular conductivities were measured [65]. The in-plane conductivity, κ_{par} improved from 18 Wcm^{-1}K^{-1} to 21.8 W cm^{-1} K^{-1} (a 21% improvement) when the concentration of ^{13}C reduced from 1.1% to 0.055%. The perpendicular conductivity, $\kappa_{\perp}$, was measured to be 26 W cm^{-1}K^{-1} with isotopic enrichment, which is a record value for CVD diamond. Analysis of the temperature dependence of κ_{par} revealed that the point-defect scattering of phonons in isotopically enriched diamond is significantly lower than expected for the natural abundance of ^{13}C and that it is responsible for the improved conductivity. The observed anisotropy $\kappa_{\text{par}}/\kappa_{\perp} = 0.84$ at room temperature is associated with the anisotropic grain structure.

7.3.2 Thermal Shock Resistance

Many high-thermal-conductivity applications where high heat loads are present require that the material exhibits good resistance to thermal shock. One such application is the use of diamond as windows and domes for high-speed aircraft and missiles. In such cases the allowable heat load depends on the parameters of windows and domes as well as heat flow regime as characterized by the Biot number, Bi = hL/κ [33]. Here h is the heat transfer coefficient at the outer surface of the window or dome and L is the characteristic dimension, which could be thickness of the window or dome. The thermal shock parameter R, which is essentially a figure of merit for comparing different materials, is defined for thermally thin (Bi < 1) and thick (Bi > 1) conditions as follows:

$$R_{\text{Thin}} = \sigma\kappa(1-\nu)/\alpha E, \quad \text{if Bi} < 1 \tag{7.3}$$

$$R_{\text{Thick}} = \sigma(1-\nu)/\alpha E, \quad \text{if Bi} > 1 \tag{7.4}$$

where σ is the flexural strength, κ is the thermal conductivity, ν is the Poisson ratio, α is the thermal expansion coefficient, and E is the elastic modulus. For

Table 7.2. Comparison of thermal shock resistance of important optical materials in thermally thin case (Bi < 1)

Material	Flexural Strength σ (MPa)	Elastic Modulus E (GPa)	Poisson Ratio ν	Thermal Conductivity κ (Wm^{-1}K^{-1})	CTE α (10^{-6} K^{-1})	Thermal Shock Parameter r_{Thin}
CVD diamond	160–400	1140	0.069	≤2300	0.8	≤939
CVD-SiC	420	466	0.21	214	2.2	69
Si$_3$N$_4$	320	310	0.27	33	1.8	14
AlN	350	440	0.29	220	4.4	28
Sapphire	400	380	0.27	24	8.8	2.1
Spinel	160–190	190	0.26	14.6	8.0	1.2–1.39
ALON	300	315	0.24	12.6	7.8	1.2
MgF$_2$	100	115	0.3	16	11.0	0.89
GaP	100	103	0.31	97	6	10.8
GaAs	60	86	0.31	53	6	4.3
CVD-ZnS	103	75	0.29	16.7	7	2.3

thermally thick case, R_{Thick} is similar to the Hasselman parameter for strong shock [66]. Table 7.2 compares the thermal shock parameter of CVD diamond with that of several competing materials [67] in the thermally thin case. We see that even though the flexural strength of CVD diamond is lower by an order of magnitude than single-crystal diamond, CVD diamond has the highest thermal shock parameter value in the thermally thin case due to its very high thermal conductivity. For the thermally thick case, however, the thermal shock parameter does not include the thermal conductivity. Consequently, CVD diamond does not perform as well and has the third highest value behind silicon nitride and silicon carbide. If the CVD processing is improved to a level so that the flexural strength of the CVD diamond is close to the single-crystal value (~3000 MPa), then the CVD diamond will have the highest value of the thermal shock parameter for the thermally thick case as well. In practice, one does not encounter many situations where the Biot number for diamond is greater than 1 because the high-thermal-conductivity diamond is usually thin (<2 mm). Thermal shock parameters specific to diamond windows and domes are discussed in more detail in Ref. [33].

7.4 High-Thermal-Conductivity Applications

The extremely high thermal conductivity of CVD diamond, along with its many other attractive properties such as low thermal expansion, high resistivity, low dielectric constant, high elastic modulus, and excellent resistance to thermal shock and chemicals, makes it an ideal candidate for a variety of applications in thermal management, optics, and electronics parts.

7.4.1 Thermal Management Applications

As power densities in electronics continue to increase with circuit miniaturization, the need to control the flow of heat has become increasingly important. Due to its very high thermal conductivity, high resistivity, and low dielectric constant, CVD diamond finds increasing use in the electronics industry for thermal management of high-power laser diodes, multichip modules and 3-dimensional architectures, and radio-frequency power amplifiers for radar and communications. Diamond helps in transporting the heat more quickly from the hot devices to the ultimate heat sink, minimizes hot spots, and lowers the junction temperatures that increase device life and reliability. In addition, diamond heat spreaders permit more closely packed circuits, which can provide faster operation without overheating. Specific examples include CVD diamond plates which can be substituted for ceramic circuit boards supporting high-power amplifiers. For large-scale cooling, multiple 10-cm diamond wafers with multichip-module attachment of chips can be stacked in close proximity to each other to achieve a short electrical delay time between wafers and still have enough thermal conductance to avoid the buildup of dangerously high temperatures. Diamond heat spreaders have increased output power and lifetimes in In-Ga-As laser diodes and diode arrays enhancing performance in industrial material processing and direct-to-plate printing systems. In power transistors operating at microwave frequencies, diamond heat spreaders provide high heat transport rates, yield higher power output with increased reliability, and can be easily metallized with appropriate patterns that can become an integral part of the transistor package design.

Another possible use of diamond is as a thin layer buried in a silicon wafer [30]; such a configuration would make use of the very low thermal boundary resistance that is possible between CVD diamond and the silicon substrate. Diamond is also being considered for use as actively cooled substrates to replace silicon in thermal management applications [20]. Microchannels are fabricated on the back side of a diamond substrate and the heat-producing device is built on its front side. Diamond conducts the heat from the device to the microchannels where the flowing coolant removes the heat. Due to its very high thermal conductivity, diamond performs much better than silicon.

7.4.2 Optics and Other Applications

Diamond is a good material for optics applications due to its high transmission in UV, visible and long-wave infrared regions, very high thermal conductivity, low value of thermal expansion coefficient, and high thermal shock resistance. Klein [68] evaluated different candidate mirror materials for cooled high-energy laser applications. He considered a 1-mm-thick face plate cooled by heat exchanger channels on the back side. The results showed that if laser beam distortions are of concern, the important parameter to consider is the thermal distortion parameter. This study ranked diamond at the top of the list of candidate mirror materials, which included CVD-SiC, Si, Mo, and Cu. Thus diamond has great potential as mirror material if issues of cost, fabrication, and scaling to large sizes can be satisfactorily resolved.

The CVD diamond can also be used in transmissive optics applications for severe environments associated with high-speed missiles, and space and laser systems. Klein [33], [67] has ranked important materials for use as windows and domes when they are suddenly exposed to a supersonic flight environment. This environment leads to intense convective heat loads due to a rise in temperature of the boundary layer. The CVD diamond ranked at the top of the list for a thermally thin case ($B_i < 1$), and third after Si_3N_4 and SiC for a thermally thick case ($B_i > 1$). Diamond is particularly suitable as a window material for the high-power industrial CO_2 lasers based on megawatt-power microwave tubes [2] due to diamond's very high thermal conductivity, good transmission at $10.6\,\mu m$, and moderate absorption coefficient. An edge-cooled diamond window transports heat to the edges quickly and thus produces smaller thermal and refractive index gradients and significantly less beam distortion than a ZnSe window. Further, the ZnSe windows cannot be adequately cooled to survive these systems due to its low thermal conductivity and low mechanical strength.

Advances made in the CVD process and diamond fabrication technologies in the last several years have demonstrated precision infrared imaging windows up to $120\,mm$ in diameter, fully polished hemispherical domes $70\,mm$ in diameter, and shallow diamond lenses for use as laser-output coupling windows [69], [70].

CVD diamond is also used for producing 1-megawatt gyrotron tubes for nuclear fusion research due to its high thermal conductivity and low loss tangent (3×10^{-5} at $144\,GHz$) [2]. These tubes measure $106\,mm$ diameter by $1.8\,mm$ thick and have water-cooled edges to keep the aperture cool. The tubes can withstand temperatures up to $450°C$ and pressures up to 7 atmospheres.

7.5 Summary and Conclusions

CVD diamond is a good material for high-heat-flux applications due to its superior thermal, optical, physical, mechanical, and electrical properties, particularly, its very high thermal conductivity. High-quality CVD diamond samples have exhibited many properties such as transmission, refractive index, thermal expansion coefficient, hardness, and thermal conductivity close to those of natural Type IIa diamond, which is the purest natural diamond. The thermal conductivity of CVD diamond depends on the particular growth method, the specific process conditions used for growth, and the typical columnar microstructure of the material. Because of the columnar microstructure, the thermal conductivity κ_{par} for heat flowing parallel to the film is smaller than the conductivity $\kappa_\perp$ for heat flowing perpendicular to the film, and this anisotropy can be explained reasonably well by the dirty-grain-boundary model. In general, high-thermal-conductivity values are obtained along the columnar grains in those areas where the grain size is large.

Due to its superior properties, CVD diamond is finding increasing use for thermal management in semiconductor and electronic devices and has exhibited considerable potential for use in high-heat-load applications associated with windows and domes for high-speed missiles and aircraft and high-energy laser mirrors for space applications. However, widespread use of diamond will require satisfactorily resolution of important issues such as its high cost, fabrication difficulties, particularly curved shape geometries, and scaling to large and uniform-thickness parts.

References

[1] Paul W. May, "CVD Diamond—a New Technology for the Future," *Endeavour Magazine* **19**, 101–6 (1995).

[2] D. C. Harris, "Review of Navy Program to Develop Optical Quality Diamond Windows and Domes," In *Proc. 9th DoD Electromagnetic Window Symposium* (2002).

[3] K. E. Spear, "Diamond—Ceramic Coating of the Future," *J. Am. Cerm. Soc.* **72**, 171–191 (1989).

[4] R. C. DeVries, "Synthesis of Diamond under Metastable Conditions," *Ann. Rev. Mater. Sci.* **17**, 161–187 (1987).

[5] B. V. Derjaguin and D. B. Fedoseev, "Synthesis of Diamond at Low Pressure," *Scientific American* **233**, 102–109 (1975).

[6] R. M. Hazen, *The Diamond Makers*, Cambridge University Press, Cambridge (1999).

[7] W. G. Eversole, "Synthesis of Diamond," *US Patent No. 3,030,187* (17 April 1962).

[8] J. C. Angus, H. A. Will, and W. S. Stanko, "Growth of Diamond Seed Crystals by Vapor Deposition," *J. Appl. Phys.* **39**, 2915 (1968).

[9] D. J. Poferi, N. C. Gardner, and J. C. Angus, "Growth of Boron Doped Diamond Seed Crystals by Vapor Deposition," *J. Appl. Phys.* **44**, 1428–34 (1973).

[10] B. V. Deryagin and D. V. Fedoseev, "Growth of Diamond and Graphite from the Gas Phase," Ch 4, Izd. Nauka, Moscow, USSR (1977).

[11] B. V. Spitsyn, L. L. Bouilov, and B. V. Deryagin, "Vapor Growth of Diamond on Diamond and other Surfaces," *J. Cryst. Growth* **52**, 219–26 (1981).

[12] S. Matsumoto, Y. Sato, M. Kamo, and N. Setaka, "Vapor Deposition of Diamond Particles from Methane," *Jpn. J. Appl. Phys.* **21**, L183–L185 (1982).

[13] M. Kamo, Y. Sato, S. Matsumoto, and N. Setaka, "Diamond Synthesis from Gas Phase in Microwave Plasma," *J. Cryst. Growth* **62**, 642–44 (1983).

[14] Y. Saito, S. Matsuda, and S. Nogita, "Synthesis of Diamond by Decomposition of Methane in Microwave Plasma," *J. Mater. Sci. Lett.* **5**, 565 (1986).

[15] S. Matsumoto, "Chemical Vapour Deposition of Diamond in RF Glow Discharge," *J. Mater. Sci. Lett.* **4**, 600–2 (1985).

[16] A. Sawabe and T. Inuzuka, "Growth of diamond thin films by electron-assisted chemical vapour deposition and their characterization," *Thin Solid Films* **137**, 89–99 (1986).

[17] R. J. Koba, "Technology of Vapor Phase Growth of Diamond Films," in *Diamond Films and Coatings*, ed. R. F. Davis (Noyes Publications, NJ, 1993), chapter 4.

[18] T. A. Grotjohn and J. Asmussen, "Microwave Plasma-Assisted Diamond Film Deposition," *Diamond Films Handbook*, eds. J. Asmussen and D. K. Reinhard (NY, Marcel Dekker, 2002), chapter 7.

[19] J. E. Yehoda, "Thermally Assisted (Hot-Filament) Deposition of Diamond," *Diamond Films Handbook*, eds. J. Asmussen and D. K. Reinhard (NY, Marcel Dekker, 2002), chapter 5.

[20] A. P. Malshe and W. D. Brown, "Diamond Heat Spreaders and Thermal Management," *Diamond Films Handbook*, eds. J. Asmussen and D. K. Reinhard (NY, Marcel Dekker, 2002), chapter 11.

[21] M. A. Prelas, G. Popovici, and L. K. Bigelow eds., *Handbook of Industrial diamonds and Diamonds Films* (New York, Marcel Dekker, 1997).

[22] S. Jin, J. E. Graebner, G. W. Kammlott, T. H. Tiefel, S. G. Kosinski, L. H. Chen, and R. A. Fastnacht, "Massive Thinning of Diamond Films by a Diffusion Process," *Appl. Phys. Lett.* **60**, 1948 (1992).

[23] S. Jin, J. E. Graebner, T. H. Tiefel, G. W. Kammlott, and G. J. Zydzik, "Polishing of CVD diamond by diffusional reaction with manganese powder," *Diamond Relat. Mater.* **1**, 949 (1992).

[24] T. B. Massalski, ed., *Binary Alloy Phase Diagrams*, 2nd ed. (Metals Park, OH: ASM, 1991), p. 860.

[25] S. Jin, J. E. Graebner, M. McCormack, T. H. Tiefel, A. Katz, and W. C. Dautremont-Smith, "Shaping of Diamond Films by Etching with Molten Rare-earth Metals," *Nature* **362**, 822 (1993).

[26] S. Jin, L. H. Chen, J. E. Graebner, M. McCormack, and M. E. Reiss, "Thermal conductivity in molten-metal-etched diamond films," *Appl. Phys. Lett* **63**, 622 (1993).

[27] S. Jin, W. Zhu, and J. E. Graebner, *Application of Diamond Films and Related materials*, eds. A. Feldman, Y. Tzeng, W. A. Yarbrough and M. Murakawa, NIST special publication 885 (Gaithersburg, MD NIST, 1995), 209.

[28] M. McCormack, S. Jin, J. E. Graebner, T. H. Tiefel, and G. W. Kammlott, "Low temperature thinning of thick chemically vapor-deposited diamond films with a molten Ce—Ni alloy," *Diamond Relat. Mater.* **3**, 254 (1994).

[29] A. Katz, K. W. Wang, F. A. Baiocchi, W. C. Dautremont-Smith, E. Lane, H. S. Luftman, R. R. Varma, and H. Curnan, "Ti/Pt/Au—Sn metallization scheme for bonding of InP-based laser diodes to chemical vapor deposited diamond submounts," *Materials Chem. Phys.* **33**, 281 (1993).

[30] M. N. Touzelbaev and K. E. Goodson, "Applications of micron-scale passive diamond layers for the integrated circuits and microelectromechanical systems industries," *Diamond Relat. Mater.* **7**, 1 (1998).

[31] H. Verhoeven, H. Reib, H. J. Fuber, and R. Zachai, "Thermal Resistance of Thin Diamond Films Deposited at Low Temperatures," *Appl. Phys. Lett.* **69**, 1562 (1996).

[32] C. A. Klein and G. Cardinale, "Young's modulus and Poisson's ratio of CVD diamond," *Diamond Relat. Mater.* **2**, 918 (1993).

[33] C. A. Klein, "Diamond Windows and Domes: Flexural Strength and Thermal Shock," *Diamond Relat. Mater.* **11**, 218 (2002).

[34] D. C. Harris, "Development of Chemical Vapor Deposited Diamond for Infrared Optical Applications, Status Report and Summary of Properties," Naval Air Warfare Center, China Lake Technical Report No. NAWCWPNS TP 8210 (July 1994).

[35] J. E. Field, *The properties of Natural and Synthetic Daiamond* (London, Academic Press, 1992) Chap 12, 13.

[36] J. A. Savage, C. J. H. Wort, C. S. J. Pickles, R. S. Sussmann, C. G. Sweeney, M. R. McClymont, J. R. Brandon, C. N. Dodge, and A. C. Beale, "Properties of Free-standing CVD Diamond Optical Components," *Proc. SPIE* **3060**, 144 (1997).

[37] Y. K. Kwon and P. Kim, "Unusually High Thermal Conductivity in Carbon Nanotubes", in *High Thermal Conductivity Materials*, eds. S. L. Shinde and J. S. Goela (Springer Verlag, 2005), Chapter 8.

[38] R. Berman, "Thermal Properties", in *The Properties of Diamond*, ed. J.E. Field (London: Academic Press, 1979), p. 3.

[39] R. Berman, P. R. W. Hudson, and M. Martinez, "Nitrogen in diamond: evidence from thermal conductivity," *J. Phys. C: solid State Phys.* **8**, L430 (1975).

[40] D. T. Morelli, T. M. Hartnett, and C. J. Robinson, "Phonon Defect Scattering in High Thermal Conductivity Diamond Films," *Appl. Phys. Lett.* **59**, 2112 (1991).

[41] C. J. Robinson, T. M. Hartnett, R. P. Miller, C. B. Willingham, J. E. Graebner, and D. T. Morelli, "Diamond for High Heat Flux Applications," *Proc. SPIE* **1739**, 146 (1993).

[42] D. T. Morelli, C. Uher, and C. J. Robinson, "Transmission of Phonons Through Grain Boundaries in Diamond Films," *Appl. Phys. Lett.* **62**, 1085 (1993).

[43] J. E. Graebner, S. Jin, G. W. Kammlott, J. A. Herb, and C. F. Gardinier, "Large Anisotropic Thermal Conductivity in Synthetic Diamond Films," *Nature* **359**, 401 (1992).

[44] J. E. Graebner, S. Jin, G. W. Kammlott, J. A. Herb, and C. F. Gardinier, "Unusually high thermal conductivity in diamond films," *Appl. Phys. Lett.* **60**, 1576 (1992).

[45] J. E. Graebner, S. Jin, J. A. Herb, and C. F. Gardinier, "Local thermal conductivity in chemical-vapor-deposited diamond," *J. Appl. Phys.* **76**, 1552 (1994).

[46] J. E. Graebner, M. E. Reiss, L. Seibles, T. M. Hartnett, R. P. Miller, and C. J. Robinson, "Phonon scattering in chemical-vapor-deposited diamond," *Phys. Rev. B* **50**, 3702 (1994).

[47] J. E. Graebner, J. A. Mucha, and F. A. Baiocchi, "Sources of thermal resistance in chemically vapor deposited diamond," *Diamond Relat. Mater.* **5**, 682 (1996).

[48] K. M. McNamara Rutledge, B. E. Scruggs, and K. K. Gleason, "Influence of hydrogenated defects and voids on the thermal conductivity of polycrystalline diamond," *J. Appl. Phys.* **77**, 1459 (1994).

[49] B. Dischler, C. Wild, W. Muller-Sebert, and P. Koidl, "Hydrogen in polycrystalline diamond: An infrared analysis," *Physica B* **185**, 217 (1993).

[50] S. Dannefaer, T. Bretagnon, and D. Kerr, "Positron lifetime investigations of diamond films," *Diamond Relat. Mater.* **2**, 1479 (1993).

[51] K. E. Goodson, O. W. Kading, and R. Zachai, "Thermal Resistances at the Boundaries of CVD Diamond Layers in Electronic Systems," *ASME HTD Proceedings* **292**, 83 (1994).

[52] K. E. Goodson, "Thermal Conduction in Nonhomogeneous CVD Diamond Layers in Electronic Microstructures," *ASME/JSME Thermal Engineering Conference* **4**, 183 (1995).

[53] J. E. Graebner, "Thermal Conductivity of Diamond Films: 0.5 μm to 0.5 mm," *Israel Journal of Chemistry* **38**, 1 (1998).

[54] K. E. Goodson, O. W. Kading, M. Rosner, and R. Zachai,, "Thermal Conduction Normal to Diamond-silicon Boundaries," *Appl. Phys. Lett.* **66**, 3134 (1995).

[55] H. Verhoeven, A. Floter, H. Reib, R. Zachai, D. Wittorf, and W. Jager, "Influence of the Microstructure on the Thermal Properties of Thin Polycrystalline Diamond Films," *Appl. Phys. Lett.* **71**, 1329 (1997).

[56] T. R. Anthony, W. F. Banholzer, J. F. Fleischer, L. Wei, P. K. Kuo, R. L. Thomas, and R. W. Pryor, "Thermal diffusivity of isotopically enriched ^{12}C diamond," *Phys. Rev. B* **42**, 1104 (1990).

[57] D. G. Onn, A. Witek, Y. Z. Qui, T. R. Anthony, and W. F. Banholzer, "Some aspects of the thermal conductivity of isotopically enriched diamond single crystals," *Phys. Rev. Lett.* **68**, 2806 (1992).

[58] J. R. Olson, R. O. Pohl, J. W. Vandersande, A. Zoltan, T. R. Anthony, and W. F. Banholzer, "Thermal conductivity of diamond between 170 and 1200 K and the isotope effect," *Phys. Rev. B* **47**, 14850 (1993).

[59] L. Wei, P. K. Kuo, R. L. Thomas, T. R. Anthony, and W. F. Banholzer, "Thermal conductivity of isotopically modified single crystal diamond," *Phys. Rev. Lett.* **70**, 3764 (1993).

[60] V. I. Nepsha, V. R. Grinberg, Yu. A. Klyuev, and A. M. Naletov, "Effect of ^{13}C isotopes on the diamond thermal conduction in the approximation of the dominant role of normal phonon-scattering processes," *Dokl. Akad. Nauk SSSR* **317**, 96 (1991).

[61] R. Berman, "Thermal conductivity of isotopically enriched diamonds," *Phys. Rev. B* **45**, 5726 (1992).

[62] K. C. Haas, M. A. Tamor, T. R. Anthony, and W. F. Banholzer, "Lattice dynamics and Raman spectra of isotopically mixed diamond," *Phys. Rev. B* **45**, 7171 (1992).

[63] Y. J. Han and P. G. Klemens, "Anharmonic thermal resistivity of dielectric crystals at low temperatures," *Phys. Rev. B* **48**, 6033 (1993).

[64] R. Berman, Thermal Conduction in Solids (Oxford Universtity Press, oxford, 1976).

[65] J. E. Graebner, T. M. Hartnett, and R. P. Miller, "Improved Thermal Conductivity in Isotopically Enriched Chemical Vapor Deposited Diamond," *Appl. Phys. Lett.* **64**, 2549 (1994).

[66] D. Hasselman, "Thermal Stress Resistance Parameters for Brittle Refractory Ceramic: a compendium," *Ceram. Bull.* **49**, 1033 (1970).

[67] C. A. Klein, "Thermal Shock Resistance of Infrared Transmitting Windows and Domes," *Opt. Eng.* **37**, 2826 (1998).

[68] C. A. Klein, "Materials for High Power Optics: Figure of Merit for Thermally Induced Beam Distortions," *Opt. Eng.* **36**, 1586 (1997).

[69] C. J. H. Wort, C. S. J. Pickles, A. C. Beale, and C. G. Sweeney, "Recent Advances in the Quality of CVD Diamond Optical Components," *SPIE Proc.* **3705**, 119 (1999).

[70] T. P. Mollart, C. J. H. Wort, C. S. J. Pickles, M. R. McClymont, N. Perkins, and K. L. Lewis, "CVD Diamond Optical Components, Multi-spectral Properties and Performance at Elevated Temperature," *Proc. SPIE* **4375**, 180 (2001).

Unusually High Thermal Conductivity in Carbon Nanotubes

Young-Kyun Kwon and Philip Kim

Recently discovered carbon nanotubes have exhibited many unique material properties including very high thermal conductivity. Strong sp^2 bonding configurations in carbon network and nearly perfect self-supporting atomic structure in nanotubes give unusually high phonon-dominated thermal conductivity along the tube axis, possibly even surpassing that of other carbon-based materials such as diamond and graphite (in plane). In this chapter, we explore theoretical and experimental investigations for the thermal-transport properties of these materials.

8.1 Introduction

The miniaturization of electrical and mechanical systems is the main achievement of modern technology, making faster and more efficient devices. With the continually decreasing size of electronic devices and microelectromechanical systems (MEMS), there is an increasing effort to use nanoscale materials as components of nanoscale devices. The thermal properties of the nanoscale materials are of fundamental interest and play a critical role in controlling the performance and stability of the device that consists of these materials. Among these materials, carbon nanotubes are of particular interest for their unique electric and thermal properties [1], [2].

Carbon nanotubes were discovered by Iijima in 1991 [3]. These novel materials, in fact, are natural extensions of fullerene clusters that have been extensively studied since the discovery of C_{60} [4] in 1985. Like other fullerenes, carbon nanotubes are made of only carbon, and the sp^2 bonding yields a π-bonding network. However, unlike most fullerenes and their derivatives, carbon nanotubes have extremely high aspect ratios. Structurally, carbon nanotubes consist of seamless cylindrical tubes that can be conceptually formed by cutting and rolling up a graphene sheet (a single layer of graphite). Single-walled nanotubes (SWNTs) consist of only a single seamless cylinder, whereas multiwalled nanotubes (MWNTs) consist of several concentric shells.

Traditionally, carbon-based materials, such as diamond and graphite, have been a material class that exhibits very high thermal conductivity. Isotope impurity-free monocrystalline diamond is one of the best thermal conductors due to the high speed of sound resulting from the stiff covalent sp^3 bonds between the carbon atoms and greatly suppressed impurity phonon scattering [5]. High thermal conductivity should also be expected in carbon nanotubes, which are held together by even stronger sp^2 bonds. These systems, consisting of seamless and atomically perfect graphitic cylinders a few nanometers in diameter, are self-supporting. Thus, the rigidity of the graphitic walls, combined with the absence of atomic defects or coupling to soft phonon modes of the embedding medium, should make isolated nanotubes very good candidates for efficient thermal conductors.

In this chapter, we present both theoretical and experimental surveys for the investigation of thermal conduction in carbon nanotubes. In the first part, we present theoretical calculations of thermal conductivity of nanotubes. After a brief discussion of phonons in carbon nanotubes, we discuss several computational methods based on molecular dynamics simulations used to determine thermal conductivity of nanotubes. The resulting thermal conductivities of carbon nanotubes and other carbon allotropes and their temperature dependencies will be reviewed. An unusually high thermal conductivity is predicted for isolated SWNTs in the calculations. In the second part, we discuss the reported experimental results in thermal-conductivity measurements of carbon nanotube materials. First, the bulk measurements including carbon nanotube composite materials are reviewed. In addition, a detailed description of recently demonstrated mesoscopic nanotube thermal-transport measurement will be presented. Finally, the comparison of theoretical and experimental results is given in the last part of the chapter, which confirms the proposed unusually high thermal conductivity in these materials.

8.2 Theory of Energy Conduction in Carbon Nanotubes

The thermal-transport properties of materials can be calculated using two main computational schemes. One scheme is the use of the Boltzmann equation, and the other is based on linear response theory from which the thermal correlation functions are derived. Whereas the former scheme, which is empirical, can be applied only to the materials that have experimental inputs available, the latter, which can be performed from first principle, is often used to predict the thermal properties of newly synthesized materials such as carbon nanotubes.

Carbon nanotubes are classified primarily into achiral and chiral nanotubes [1], [6]. An achiral nanotube exhibits a mirror symmetry on the plane normal to the tube axis whereas a chiral one shows a spiral symmetry. There are only two types of achiral nanotubes that show higher symmetry than chiral tubes. One is an "armchair" type and the other a "zigzag," as discussed later. The structure of a nanotube is more specified by the orientation of hexagonal carbon rings on cylindrical graphene sheets with respect to the tube axis.

This orientation is characterized by the chiral index (n, m) defined by the chiral vector $\mathbf{C}_h$

$$\mathbf{C}_h = n\mathbf{a}_1 + m\mathbf{a}_2, \qquad (8.1)$$

where $\mathbf{a}_i (i = 1, 2)$ are real space unit vectors of the hexagonal lattice. This chiral vector, as shown in Fig. 8.1, connects two equivalent sites O and A on a graphene sheet. Its magnitude $\mathbf{C}_h$ represents a circumferential length of a nanotube being characterized by $\mathbf{C}_h$. The direction perpendicular to $\mathbf{C}_h$ becomes a tube axis. A pair of integers (n, m) in Eq. (8.1), specifying all possible chiral vectors, defines a different way of rolling the graphene sheet to form a nanotube. Zigzag nanotubes, which have the zigzag shape of the cross-sectional ring, and armchair nanotubes, which have the armchair shape, are denoted by the vectors $(n, 0)$ and (n, n), respectively.

The tube diameter d_t is given by

$$d_t = \mathbf{C}_h/\pi$$
$$= \sqrt{3}d_{CC}(n^2 + nm + m^2)^{1/2}/\pi, \qquad (8.2)$$

where d_{CC} is the nearest-neighbor distance between two carbon atoms (in graphite $d_{CC} = 1.42\text{Å}$). And the chiral angle θ, defined as the angle between the chiral vector $\mathbf{C}_h$ and the lattice vector $\mathbf{a}_1$, is given by

$$\cos\theta = \frac{\mathbf{C}_h \cdot \mathbf{a}_1}{\mathbf{C}_h a}$$
$$= \frac{2n + m}{2\sqrt{n^2 + nm + m^2}}. \qquad (8.3)$$

The chiral angle θ is just in the range of $0 \le |\theta| \le 30°$, because of the hexagonal symmetry of the graphene sheet. Armchair nanotubes, in particular, correspond to $\theta = 30°$ and zigzag ones $\theta = 0°$.

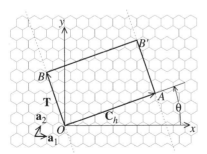

Fig. 8.1. A nanotube can be constructed by connecting site O to site A and site B to site B'. This nanotube is $(6, 3)$ (see the text for the tube classification). The chiral vector $\mathbf{C}_h$ and the translational vector $\mathbf{T}$ of the nanotube are represented by arrow lines of OA and OB, respectively. The rectangle $OAB'B$ defines the unit cell of the nanotube.

The vector $\mathbf{T}$, called the translation vector, is parallel to the tube axis, that is, perpendicular to the chiral vector $\mathbf{C}_h$. Using the relation of $\mathbf{C}_h \cdot \mathbf{T} = 0$, the vector $\mathbf{T}$, which becomes the lattice vector in a 1D tube unit cell, can be expressed in terms of the basis vectors $\mathbf{a}_i$ as

$$\mathbf{T} = \frac{1}{d_R}[-(n + 2m)\mathbf{a}_1 + (2n + m)\mathbf{a}_2], \tag{8.4}$$

where d_R is the greatest common divisor of $(n + 2m)$ and $(2n + m)$. Furthermore, d_R can be expressed in terms of the greatest common divisor d of n and m. If $n - m$ is a multiple of $3d$, $d_R = 3d$; otherwise $d_R = d$. Note that the bigger d_R, the smaller the length of $\mathbf{T}$. For example, $\mathbf{T} = -\mathbf{a}_1 + \mathbf{a}_2$ for any (n, n) armchair nanotubes ($d_R = 3d = 3n$) and $\mathbf{T} = -\mathbf{a}_1 + 2\mathbf{a}_2$ for any $(n, 0)$ zigzag nanotubes ($d_R = n$). A $(6, 3)$ nanotube ($d = d_R = 3$) shown in Fig. 8.1 has $\mathbf{T} = -4\mathbf{a}_1 + 5\mathbf{a}_2$.

The rectangle formed by two vectors $\mathbf{C}_h$ and $\mathbf{T}$ determines the unit cell of a nanotube, as shown as $OAB'B$ in Fig. 8.1. Because, in this unit cell there are $2N$ carbon atoms (the number of hexagons N in the unit cell is expressed in terms of n, m, and d_R as $N = 2(n^2 + nm + m^2)/d_R$) we expect there are $6N$ phonon-dispersion branches.

The corresponding vectors in reciprocal space are determined from the relations

$$\mathbf{C}_h \cdot \mathbf{K}_1 = 2\pi, \quad \mathbf{T} \cdot \mathbf{K}_1 = 0,$$
$$\mathbf{C}_h \cdot \mathbf{K}_2 = 0, \quad \mathbf{T} \cdot \mathbf{K}_2 = 2\pi, \tag{8.5}$$

where $\mathbf{K}_1$ is in the circumferential direction and $\mathbf{K}_2$ along the tube axis. The resulting expressions for $\mathbf{K}_1$ and $\mathbf{K}_2$ are given by

$$\mathbf{K}_1 = \frac{1}{N}(-t_2\mathbf{b}_1 + t_1\mathbf{b}_2),$$
$$\mathbf{K}_2 = \frac{1}{N}(m\mathbf{b}_1 - n\mathbf{b}_2), \tag{8.6}$$

where $\mathbf{b}_i$ ($i = 1, 2$) are the reciprocal lattice vectors of the hexagonal lattice. $\mathbf{K}_2$ is the *reciprocal lattice vector* that is the counterpart of $\mathbf{T}$ in real space, whereas $\mathbf{K}_1$ is just a corresponding vector to $\mathbf{C}_h$, which gives discrete k values in the circumferential direction.

In this section, we discuss the computational approach to probe the thermal properties of carbon nanotubes. After we review phonon-dispersion relations of 2-dimensional (2D) graphite and carbon nanotubes, we describe computational methods based on molecular dynamics simulations for determining thermal conductivity and their pitfalls when applied to nanotubes. Combining equilibrium and nonequilibrium molecular dynamics simulations with Green-Kubo formalism, we determine the thermal conductivity of SWNTs and other carbon allotropes. Our results suggest an unusually high value $\kappa \approx 6{,}600\,\mathrm{W/m \cdot K}$ for an isolated $(10, 10)$ nanotube at room temperature, comparable to the thermal conductivity of a hypothetical isolated graphite monolayer or diamond. We find that these high values of κ are associated with large phonon mean free paths in these systems.

8.2.1 Phonons in Carbon Nanotubes

It has been shown that the 1-dimensional (1D) electronic band structure of carbon nanotubes [7], [8], [9] can be obtained from that of an ideal 2D graphene sheet using the zone-folding approach. Likewise, the zone-folding approach has been used to determine the phonon-dispersion relations of carbon nanotubes from those of the graphene sheet [6], [10], [11], obtained by solving the secular equation of its dynamical matrix to be determined by a simple force constant model. (See [6], [11] for a more detailed description of phonon modes of 2D graphite and carbon nanotubes.)

The equations of motion of the lattice are, in general, expressed as

$$m_i \ddot{\mathbf{x}}_i = -\sum_j K^{(ij)}(\mathbf{x}_i - \mathbf{x}_j), \quad (i = 1, \ldots, N), \tag{8.7}$$

where m_i and $\mathbf{x}_i$ are, respectively, the mass and the displacement vector from its equilibrium position of the ith atom among N atoms in a unit cell, and $K^{(ij)}$ is the 3×3 force constant matrix between the ith and jth atoms. $\sum_j$ means the summation taken over all interacting neighbor atoms, which are usually considered up to the nth nearest neighbors, including ones in other unit cells. If we seek normal mode solutions of Eq. (8.7)

$$\mathbf{x}_l = \sum_{\mathbf{k}} \mathbf{u}_{\mathbf{k}}^{(l)} e^{-i(\mathbf{k} \cdot \mathbf{r}_l - \omega t)}, \tag{8.8}$$

where the summation is taken over all the wave vectors $\mathbf{k}$ in the first Brillouin zone. Here $\mathbf{r}_l$ is the equilibrium position of the lth atom and $\mathbf{u}_{\mathbf{k}}^{(l)}$ denotes the Fourier coefficient of $\mathbf{x}_l$. We assume the same eigenfrequencies ω for all $\mathbf{x}_l$. Substituting Eq. (8.8) into Eq. (8.7) and using the orthogonal condition in reciprocal space,

$$\sum_{\mathbf{r}_l} e^{-i(\mathbf{k}-\mathbf{k}') \cdot \mathbf{r}_l} = \delta(\mathbf{k} - \mathbf{k}'), \tag{8.9}$$

where $\delta(\mathbf{k} - \mathbf{k}')$ is a delta function in the continuum k space, Eq. (8.7) becomes

$$-m_i \omega^2 I \mathbf{u}_{\mathbf{k}}^{(i)} = -\sum_j K^{(ij)} \left(\mathbf{u}_{\mathbf{k}}^{(i)} - e^{-i\mathbf{k} \cdot \mathbf{r}_{ij}} \mathbf{u}_{\mathbf{k}}^{(j)} \right), \quad (i = 1, \ldots, N), \tag{8.10}$$

where $\mathbf{r}_{ij} = \mathbf{r}_i - \mathbf{r}_j$ and I is a 3×3 identity matrix. This equation can be written more compactly in a tensor form as

$$\mathcal{D}(\mathbf{k})\mathbf{u}_{\mathbf{k}} = 0, \tag{8.11}$$

where $\mathcal{D}(\mathbf{k})$ is a $3N \times 3N$ matrix called a dynamical matrix and decomposed into the total N^2 number of 3×3 submatrices $\{\mathcal{D}^{(ij)}(\mathbf{k})\}$ expressed as

$$\mathcal{D}^{(ij)}(\mathbf{k}) = \left(\sum_l K^{(il)} - m_i \omega^2(\mathbf{k}) I \right) \delta_{ij} - K^{(ij)} e^{i\mathbf{k}\cdot\mathbf{r}_{ij}}. \qquad (8.12)$$

Note that the dynamical matrix includes the contributions from all interacting neighbor atoms. Equation (8.11) is simply an eigenvalue equation, whose nontrivial solutions are obtained by finding the eigenvalues $\omega^2(\mathbf{k})$ resulting when the secular equation $\det \mathcal{D}(\mathbf{k}) = 0$ is solved for a given $\mathbf{k}$ vector.

For a single graphene sheet, in the unit cell of which there are two carbon atoms, α and β, its dynamical matrix will be a 6×6 matrix and can be expressed as

$$\begin{pmatrix} \mathcal{D}^{(\alpha\alpha)} & \mathcal{D}^{(\alpha\beta)} \\ \mathcal{D}^{(\beta\alpha)} & \mathcal{D}^{(\beta\beta)} \end{pmatrix} \qquad (8.13)$$

in terms of the 3×3 submatrices $\mathcal{D}^{(ij)}$, $(i, j = \alpha, \beta)$. $\mathcal{D}^{(ij)}$ contains all the contributions to the atom i from up to the fourth nearest-neighbor atoms equivalent to the atom j. As given in Eq. (8.12), $\mathcal{D}^{(ij)}$ is constructed by calculating the force constant matrix $K^{(ij)}$, which is composed of the force constant parameters $f_r^{(n)}$, $f_{t_i}^{(n)}$, and $f_{t_o}^{(n)}$, $(n = 1, \ldots, 4)$, which are determined by the interactions of the nth neighbor atoms, in the radial (bond stretching), in-plane and out-of-plane tangential (bond-bending) directions, respectively. Table 8.1, which was originally shown in [10], gives values for the force constant parameters for 2D graphite obtained by fitting to experimental phonon-dispersion relations measured along the ΓM direction [12], [13].

Consider the contributions to the atom α from three first nearest neighbor atoms $\beta_1^{(1)}$, $\beta_2^{(1)}$, and $\beta_3^{(1)}$. Assume that the atom $\beta_1^{(1)}$ is in the same unit cell as the atom α, whereas each of the other two atoms is in a neighboring unit cell. In a coordinate system in which the atom α is at the origin, the atom β_1 is on the x-axis, and the z-axis perpendicular to the graphene sheet is passing by the atom α. Then, the contribution from the atom $\beta_1^{(1)}$ to the

Table 8.1. Values for the force constant parameters $f_r^{(n)}$, $f_{t_i}^{(n)}$, and $f_{t_o}^{(n)}$ $(n = 1, \ldots, 4)$, for 2D graphite up to the $n =$ fourth nearest neighbor interactions, originally shown in [10]. The subscripts r, t_i, and t_o refer to radial, tangential in plane, and out of plane, respectively. The values are given in units of 10^4 dyn/cm.

Radial		Tangential	
$f_r^{(1)} = 36.50$	$f_{t_i}^{(1)} = 24.50$	$f_{t_o}^{(1)} = 9.82$	
$f_r^{(2)} = 8.80$	$f_{t_i}^{(2)} = -3.23$	$f_{t_o}^{(2)} = -0.40$	
$f_r^{(3)} = 3.00$	$f_{t_i}^{(3)} = -5.25$	$f_{t_o}^{(3)} = 0.15$	
$f_r^{(4)} = -1.92$	$f_{t_i}^{(4)} = 2.29$	$f_{t_o}^{(4)} = -0.58$	

force constant matrix $K^{(\alpha\beta)}$ is given by

$$\tilde{K}\left(\alpha\beta_1^{(1)}\right) = \begin{pmatrix} f_r^{(1)} & 0 & 0 \\ 0 & f_{t_i}^{(1)} & 0 \\ 0 & 0 & f_{t_o}^{(1)} \end{pmatrix}. \tag{8.14}$$

The force constant matrix K is expressed in terms of $\tilde{K}$ as

$$K^{(\alpha\beta)} \equiv \sum_n \sum_l \tilde{K}\left(\alpha\beta_l^{(n)}\right). \tag{8.15}$$

Using a rotation matrix R_l given by

$$R_l = \begin{pmatrix} \cos\phi_l & \sin\phi_l & 0 \\ -\sin\phi_l & \cos\phi_l & 0 \\ 0 & 0 & 1 \end{pmatrix}, \quad (l = 2, 3), \tag{8.16}$$

where $\phi_l = 2(l - 1)\pi/3$ are rotational angles between $\beta_1^{(1)}$ and $\beta_l^{(1)}$, the contributions from the other two atoms $\beta_2^{(1)}$ and $\beta_3^{(1)}$, both of which are geometrically equivalent to the atom $\beta_1^{(1)}$, are expressed by

$$\tilde{K}\left(\alpha\beta_l^{(1)}\right) = R_l^{-1}\tilde{K}\left(\alpha\beta_1^{(1)}\right)R_l, \quad (l = 2, 3). \tag{8.17}$$

We also calculate all other contributions of the nth nearest neighbor atoms to the force constant matrix in a similar way. Considering all these contributions and the corresponding phase factor $e^{i\mathbf{k}\cdot\mathbf{r}_{ij}}$, we complete the construction of the dynamical matrix $\mathcal{D}(\mathbf{k})$ and thus obtain the phonon-dispersion relations by solving the secular equation $\det \mathcal{D}(\mathbf{k}) = 0$.

The phonon-dispersion relations for 2D graphite are displayed in Fig. 8.2. There are a total of six phonon branches, three of which correspond to acoustic modes that have zero energy at the Γ point, and the other three are optical. The lowest mode in energy corresponds to the t_o mode, and the second and third lowest ones to t_i and r, respectively, near the Γ point. The t_i and r modes, both of which are *in-plane* modes, show a linear k dependence as usually seen for acoustic modes, whereas the t_o mode that is an *out-of-plane* mode, shows a special k^2 dependence, which comes from the three-fold rotational symmetry (C_3) around the z-axis. Although no linear combination of k_x and k_y can be invariant under C_3 rotation, the quadratic form of $k_x^2 + k_y^2$ as well as a constant is invariant. Similarly, the optical t_o mode ($\omega \sim 865\,\text{cm}^{-1}$ at $k = 0$) shows a k^2 dependence.

The phonon-dispersion relations for SWNTs are obtained by the zone-folding method from those of 2D graphite $\omega_{\text{gra}}^l(\mathbf{k})$ (see Fig. 8.2), where $l = 1, \ldots, 6$ labels six phonon branches and $\mathbf{k}$ is a vector in 2D reciprocal space. Supposed that an SWNT has N hexagons in its unit cell, that is, $2N$ carbon

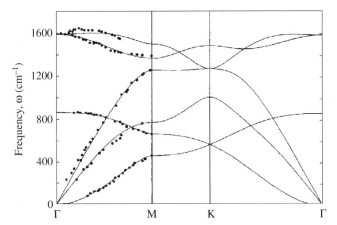

Fig. 8.2. The phonon-dispersion relations for the 2D graphene sheet, plotted along high-symmetry directions, using the set of the force constant parameters in Table 8.1 [10], [11]. The points shown along the ΓM line are experimental values [12], [13]. [Courtesy of Millie Dresselhaus]

atoms. Then its phonon-dispersion relations $\omega_{nt}^{l\nu}(k)$ are given in terms of $\mathbf{K}_i$, $(i = 1, 2)$ given in Eq. (8.6) by

$$\omega_{nt}^{l\nu}(k) = \omega_{gra}^l \left(\nu \mathbf{K}_1 + k \frac{\mathbf{K}_2}{|\mathbf{K}_2|} \right), \tag{8.18}$$

where $\nu = 0, \ldots, N - 1$ and k is a 1D wave vector of the nanotube ranging from $-\pi/T$ to π/T with $T = |\mathbf{T}|$.

Although the zone-folding method describes the overall features of the phonon modes of a nanotube quite well, some of their features, especially near the low-frequency region, are not described by the method. For example, the acoustic t_o mode of 2D graphite, which has $\omega = 0$ at $\mathbf{k} = 0$, corresponds to a radial breathing mode in the carbon nanotube, which has *nonzero* frequency, $\omega > 0$, at the Γ point. Another example is an acoustic mode of a nanotube, the vibration of which is normal to the nanotube axis corresponding a rigid shift of the nanotube along, for example, the x-axis. This mode is formed by a linear combination of in-plane and out-of-plane modes in 2D graphite in which one mode is not coupled with the other. Hence, this mode is a unique acoustic mode of the nanotube.

To avoid considering these additional physical concepts to determine the phonon-dispersion relations of nanotubes, we solve the $6N \times 6N$ dynamical matrix $\mathcal{D}(\mathbf{k})$ of a nanotube, which has $2N$ carbon atoms denoted by α_i and β_j $(i, j = 1, \ldots, N)$, directly. (N atoms of α_i, (or β_j) are geometrically equivalent to each other.) Similarly as discussed earlier $\mathcal{D}$ is decomposed into 3×3 small matrices $\mathcal{D}^{(\alpha_i \beta_j)}$ for a pair of α_i and β_j atoms as

$$
\mathcal{D} = \begin{pmatrix} \ddots & & & \\ & \mathcal{D}^{(\alpha_i \alpha_j)} & \cdots & \mathcal{D}^{(\alpha_i \beta_j)} & \\ & \vdots & \ddots & \vdots & \\ & \mathcal{D}^{(\beta_i \alpha_j)} & \cdots & \mathcal{D}^{(\beta_i \beta_j)} & \\ & & & & \ddots \end{pmatrix} . \tag{8.19}
$$

By a similar discussion on unitary transformation (see (8.14)–(8.17)), a contribution of the force constant matrix, $\tilde{K}^{(\alpha_i \beta_j^{(n)})}$ is calculated in terms of the nonzero contribution of the force constant matrix related to α_1 or β_1, using, for example,

$$
\tilde{K}^{\left(\alpha_i \beta_j^{(n)}\right)} = R_{i-1}^{-1} \tilde{K}^{\left(\alpha_1 \beta_{j-i+1}^{(n)}\right)} R_{i-1}, \tag{8.20}
$$

where R_l is a unitary matrix for rotation by an angle $\phi = 2l\pi/N$ around the tube axis and given by Eq. (8.16) if the z-axis is taken for the tube axis. $\tilde{K}^{(\alpha_1 \beta_1^{(1)})}$, in which $\beta_1^{(1)}$ is a nearest neighbor atom of the atom α_1 that is on the x-axis, is determined by rotating the matrix of Eq. (8.14) by an angle $\pi/6 - \theta$ around the x-axis, and then by an angle $\xi/2$ around the z-axis. Here, θ is the chiral angle of the tube and ξ is the angle between α_1 and $\beta_1^{(1)}$ around the z-axis. Once $\tilde{K}^{(\alpha_1 \beta_1^{(1)})}$ is determined, all other contributions to the force constant matrices $\tilde{K}^{(\alpha_i \beta_j^{(1)})}$, $(i, j = 1, \ldots, N)$ of the nearest atoms β_j are obtained using Eq. (8.20). Similarly, the other contributions from nth nearest neighbor atoms are also determined, as are all of the force-constant matrices. The force-constant tensor along with the corresponding phase factor $e^{ikz_{ij}}$, where z_{ij} is the z-component of $\mathbf{r}_{ij}$, determines the dynamical matrix $\mathcal{D}$ to be solved for the phonon-dispersion relations of the nanotube. (For a more detailed description of the dynamical matrix $\mathcal{D}$ and consideration of curvature effects on the force-constant parameters, see [6]).

For example, the phonon-dispersion relations for a $(10, 10)$ carbon nanotube, which has $2N = 40$ carbon atoms in its unit cell, are displayed in Fig. 8.3. For 40 atoms, there should be 120 $(= 3 \times 2N)$ vibrational degrees of freedom, but they exhibit only 66 distinct phonon modes, because 54 phonon branches are doubly degenerate and 12 modes are nondegenerate. The corresponding density of states (DOS) for a $(10, 10)$ nanotube is shown in Fig. 8.4, where, for comparison, that for a 2D graphene sheet is also shown with the same units of state/(C atom)/cm^{-1} scaled by a factor. The overall feature for the nanotube is similar to that for the graphene, because the phonon-dispersion relations of the former are related to the zone-folding of the latter. The differences between their detailed features originate from the van Hove singularities existing only in 1D nanotubes and the low-energy modes near the Γ point, as discussed later.

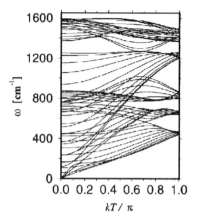

Fig. 8.3. The phonon-dispersion relations for a $(10, 10)$ carbon nanotube [11]. There are $2N = 40$ carbon atoms in its unit cell, thus 120 vibrational degrees of freedom, but there are only 66 distinct phonon modes because of degeneracies. The wave vector k is given in units of π/T. [Courtesy of Millie Dresselhaus]

Now, we consider the low-energy phonon modes near the Γ point. Shown in Fig. 8.5 are four acoustic branches, that have zero energy $(\omega = 0)$ at the Γ point $(k = 0)$, showing a linear k dependence. The transverse acoustic modes, which are doubly degenerated at the lowest energy curve, result from the vibrations (along the x- or y-direction) perpendicular to the tube axis. These modes do not exist in 2D graphite, as discussed earlier. The longitudinal acoustic mode or the vibration in the direction of the tube axis (z-axis) is shown as the highest energy acoustic mode in Fig. 8.5. The other acoustic phonon curve, located between the lowest and highest ones in Fig. 8.5,

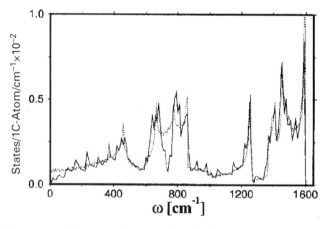

Fig. 8.4. The phonon densities of states for a $(10, 10)$ carbon nanotube (solid line) and a graphene sheet (dotted line) [11]. [Courtesy of Millie Dresselhaus]

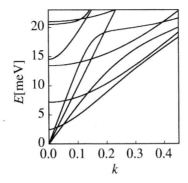

Fig. 8.5. The phonon-dispersion relations for a $(10, 10)$ carbon nanotube in a low-energy region near the Γ point [6], [11]. Shown are four acoustic phonon modes (two degenerate TA modes, one twist mode, and one LA mode, listed in order of increasing energy, respectively) and several lowest optical subbands including the radial breathing mode at $\omega(k = 0) \approx 20.5$ meV. [Courtesy of Millie Dresselhaus]

is related to a rigid rotation of the tube around the tube axis. This mode, corresponding to the in-plane transverse acoustic mode for 2D graphite, is called the twisting mode. The sound velocities of the transverse, twisting, and longitudinal acoustic modes for a $(10, 10)$ carbon nanotube are estimated to be $v_{TA} = 9.43 \times 10^3$ m/s, $v_{TW} = 15.0 \times 10^3$ m/s, and $v_{LA} = 20.35 \times 10^3$ m/s, respectively.

Several lowest optical subbands obtained by the zone-folding of 2D graphite, are also shown in Fig. 8.5. Among them are included an E_{2g} mode at $\omega(k = 0) = \sim 17$ cm$^{-1} = 2.1$ meV, an E_{1g} mode at ~ 118 cm$^{-1} = 14.6$ meV, and an A_{1g} mode at ~ 165 cm$^{-1} = 20.5$ meV, which is the radial breathing mode corresponding to the *acoustic* mode showing a k^2 dependence in a graphene sheet. Some phonon bands, which have the same symmetry, show anticrossing behavior because they couple to each other, whereas the modes with different symmetries simply cross because they do not interact with each other.

8.2.2 Computational Methods

Here we describe several computational methods that calculate the thermal conductivity and its temperature dependence. The thermal conductivity tensor $\mathbf{\Lambda}$ is related to the thermal current density $\mathbf{J}$ and the temperature gradient ∇T by

$$\mathbf{J} = -\mathbf{\Lambda}\nabla T, \qquad (8.21)$$

known as Fourier's definition of the thermal conductivity. If we consider the thermal conduction of a solid along a particular direction, for example, along the z-axis, the thermal conductivity κ can be expressed by combining

Eq. (8.21) with the continuity equation for heat conduction as

$$\frac{1}{A}\frac{dQ}{dt} = -\kappa\frac{dT}{dz},$$ (8.22)

where dT/dz is the z-component of the temperature gradient and dQ is the heat flowing along the z-axis through the cross-sectional area A during the time interval dt.

Electrons and phonons, in principle, contribute to thermal conduction in solids. Both theoretical [14] and experimental studies [15], [16] show that the dominant contribution to heat conductance in graphite and nanotubes comes from phonons, whereas the contribution from electrons is extremely small even at low temperatures. In the following, we consider only the phonon contribution to thermal conduction in nanotubes.

The thermal conductivity κ is proportional to $\sum Cvl$, where C is the specific heat capacity, v the speed of sound, l the phonon mean free path, and $\sum$ means the summation over all phonon modes described in Sect. 8.2.1. The mean free path l can, in general, be determined from two contributions to the inelastic phonon scattering processes, which can be expressed as

$$\frac{1}{l} = \frac{1}{l_{st}} + \frac{1}{l_{Um}}.$$ (8.23)

where the first term l_{st} on the right-hand side is limited by scattering from sample boundaries (related to grain sizes) and defects, which are dominant at low temperatures. l_{st} is, therefore, independent of temperature so that the thermal conductivity κ shows its temperature dependence similar to that of the specific heat, C at low temperatures. On the other hand, the l_{Um} is determined by phonon-phonon Umklapp scattering that is dominant at high temperatures. As the temperature increases, the Umklapp scattering becomes more frequent, and thus it reduces l_{Um} further. At high temperatures, therefore, the temperature dependence of κ is expected to be dominated by l_{Um}. Such strong dependence of the thermal conductivity κ on l is demonstrated by the reported thermal conductivity values in the basal plane of graphite, which scatter by nearly two orders of magnitude [17]. Most factors determining the mean free path l of a sample, such as its grain size, sample quality, and isotope ratio, vary sample by sample.

We describe a few methods based on molecular dynamics simulations that have been used to calculate thermal conductivity of nanotubes.

8.2.2.1 Direct Molecular Dynamics Approach Based on Velocity Rescaling. To determine the thermal conductivity of nanotubes, we first used a method based on a direct molecular dynamics simulation, which had been successfully applied for glasses [18]. As illustrated schematically in Fig. 8.6(a), we consider a periodic array of hot and cold plates (e.g., a single circumferential ring in a nanotube) perpendicular to a direction (e.g., tube axis, z) along

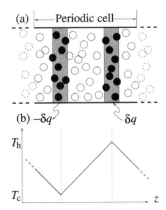

Fig. 8.6. (a) Schematic diagram of the direct molecular dynamics approach used to determine the thermal conductivity κ. Two gray regions represent the hot and cold plates maintained by subtracting δq from the energy of the particles in the cold one and adding it to that in the hot one. The particles in these two regions are depicted by a black solid circle. Through other particles depicted by an empty circle, heat gets transferred from the hot plate to the cold plate. (b) Ideal linear temperature profile along the z-axis, along which κ is being calculated.

which κ will be calculated. Heat exchange δq with hot and cold plates is achieved by velocity rescaling of particles in the plates. Due to the velocity rescaling, heat transfer is imposed, that is, thermal current flows between two regions. Once the system reaches a steady current state, one can determine thermal current density J_z along the z-direction to be $\delta q/(2A\Delta t)$, where A is a cross-sectional area of the hot plate and Δt is a time step for the molecular dynamic simulation. The thermal conductivity, then, can be calculated using Eq. (8.21), or

$$\kappa = -\frac{J_z}{\partial T/\partial z}, \tag{8.24}$$

when the z-component of the temperature gradient, $\partial T/\partial z$, does not vary significantly, as shown in Fig. 8.6(b), which exhibits an *ideal* linear temperature profile or a constant temperature gradient.

A real molecular dynamics simulation applied to a (10, 10) carbon nanotube, however, reveals a significant deviation from a stable linear temperature profile, as displayed in Fig. 8.7. In addition, the perturbations imposed by the heat transfer limit the effective phonon mean free path artificially below the unit cell size. Because the unit cell sizes tractable in our molecular dynamics simulations are significantly smaller than the phonon mean free path l of nanotubes, it has been found to be difficult to achieve the convergence of the simulations. Due to their high degree of long-range order, nanotubes exhibit an unusually long phonon mean free path l over hundreds of nanometers.

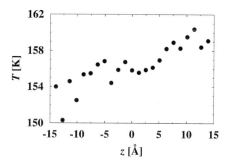

Fig. 8.7. Temperature profile of a (10, 10) nanotube along the tube axis.

8.2.2.2 Equilibrium Molecular Dynamics Simulations Based on the Green-Kubo Formalism. As an alternative approach determining the thermal conductivity, we used equilibrium molecular dynamics simulations [19], [20] based on the Green-Kubo relation for the Navier-Stokes thermal conductivity coefficient, which is derived in a relatively straightforward way from the Langevin equation [21]. The Green-Kubo expression relates the thermal conductivity to the integral over time t of the heat flux autocorrelation function by [21], [22]:

$$\kappa = \frac{1}{3Vk_BT^2} \int_0^\infty C(t)dt, \tag{8.25}$$

where k_B is the Boltzmann constant, V is the volume, T is the temperature of the sample, and $C(t)$ is the the heat-flux autocorrelation function given by

$$C(t) = \langle \mathbf{J}(t) \cdot \mathbf{J}(0) \rangle, \tag{8.26}$$

where the angled brackets $\langle \cdots \rangle$ denote an ensemble average. (It is usually very difficult to evaluate $C(t)$ quantum mechanically. Considering the quantum effects and the anharmonicity in the interaction potential, it has been proved that the classical autocorrelation function can be used with validity to calculate the thermal conductivity [23].) The heat-flux vector $\mathbf{J}(t)$ is defined by

$$\begin{aligned} \mathbf{J}(t) &= \frac{d}{dt} \sum_i \mathbf{r}_i \Delta e_i \\ &= \sum_i \mathbf{v}_i \Delta e_i + \sum_i \sum_{j(\neq i)} \mathbf{r}_{ij}(\mathbf{f}_{ij} \cdot \mathbf{v}_i), \end{aligned} \tag{8.27}$$

where $\mathbf{v}_i$ is the velocity of atom i and $\Delta e_i = e_i - \langle e \rangle$, the excess energy of atom i with respect to the average energy per atom $\langle e \rangle$. $\mathbf{r}_i$ is its position and $\mathbf{r}_{ij} = \mathbf{r}_i - \mathbf{r}_j$. Assuming that the total potential energy $U = \sum_i u_i$ can be expressed as a sum of binding energies u_i of individual atoms, $\mathbf{f}_{ij} = -\nabla_i u_j$, where ∇_i is the gradient with respect to the position of atom i.

Once $\mathbf{J}(t)$ is known, the thermal conductivity can be calculated using Eqs. (8.25) and (8.26). We found, however, that these results depend sensitively on the initial conditions of each simulation, thus necessitating a large ensemble of simulations. This high computational demand was further increased by the slow convergence of the autocorrelation function, requiring long integration time periods. This is illustrated in Fig. 8.8, which shows the autocorrelation function of a (10, 10) carbon nanotube as a function of time. The convergence has not been achieved even after several tens of thousands of molecular dynamics time steps of $\Delta t = 5 \times 10^{-16}$ sec. Moreover, because the autocorrelation function represents the average response to the fluctuation of the *equilibrium* system, which is fairly small, the signal-to-noise ratio is often small.

8.2.2.3 Nonequilibrium Molecular Dynamics Simulations Based on the Green-Kubo Formalism. To overcome these disadvantages, we now introduce an alternative approach [24] that uses molecular dynamics simulations based on *nonequilibrium* thermodynamics [25], [26]. It has been shown that this approach, developed in a computationally efficient manner [27], reduces the inefficiencies that occur in equilibrium approach. In the following, we describe briefly the nonequilibrium molecular dynamics simulations combined with the Green-Kubo formalism.

In this approach, the temperature T of the sample is regulated by a Nosé-Hoover thermostat [28], [29], which indicates the temperature of a surrounding thermal reservoir. An important fact that makes this approach *nonequilibrium* is the introduction of a small fictitious "thermal force," which improves the signal-to-noise level of the response dramatically. The fictitious thermal force $\mathbf{F}_e$, which has a dimension of *inverse length*, is equally applied to individual atoms. This fictitious force $\mathbf{F}_e$ and the Nosé-Hoover thermostat impose an additional force $\Delta\mathbf{F}_i$ on each atom i. This additional force modifies

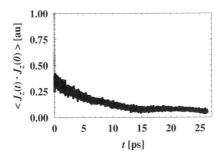

Fig. 8.8. Autocorrelation function calculated from the z-component of the heat flux vector, $J_z(t)$ for a $(10, 10)$ carbon nanotube.

the gradient of the potential energy and is given by

$$\Delta \mathbf{F}_i = \Delta e_i \mathbf{F}_e - \sum_{j(\neq i)} \mathbf{f}_{ij}(\mathbf{r}_{ij} \cdot \mathbf{F}_e)$$

$$+ \frac{1}{N} \sum_j \sum_{k(\neq j)} \mathbf{f}_{jk}(\mathbf{r}_{jk} \cdot \mathbf{F}_e) - \alpha \mathbf{p}_i. \tag{8.28}$$

Here, α is the Nosé-Hoover thermostat multiplier acting on the momentum $\mathbf{p}_i$ of atom i. α is calculated using the time integral of the difference between the instantaneous kinetic temperature T of the system and the heat bath temperature T_{eq}, from $\dot{\alpha} = (T - T_{eq})/Q$, where Q is the thermal inertia. The third term in Eq. (8.28) guarantees that the net force acting on the entire N-atom system vanishes. With the additional force $\Delta \mathbf{F}_i$ for a given value of $\mathbf{F}_e$, the heat-flux vector $\mathbf{J}(\mathbf{F}_e, t)$ is determined, for a given time t, using Eq. (8.28). The resulting thermal conductivity along the z-axis is given by

$$\kappa = \lim_{\mathbf{F}_e \to 0} \lim_{t \to \infty} \frac{\langle J_z(\mathbf{F}_e, t) \rangle}{F_e T V}, \tag{8.29}$$

where $J_z(\mathbf{F}_e, t)$ is the z-component of the heat-flux vector for a particular time t and V is the volume of the sample.

In low-dimensional systems, such as nanotubes or graphene monolayers, we infer the volume from the way these systems pack in space to convert thermal conductance of a system to thermal conductivity of a material. (Nanotubes form bundles and graphite forms a layered structure, both with an inter-wall separation of $\approx 3.4\,\text{Å}$.)

8.2.3 Thermal Conductivity of Carbon Nanotubes

We now present the results of nonequilibrium molecular dynamics simulations combined with the Green-Kubo formalism described in Sect. 8.2.2. We have used the Tersoff potential [30], [31], which has been augmented by van der Waals interactions fitted from interlayer interactions in graphite [32], for atomic interactions in the molecular dynamics simulations. The temperature dependence of the thermal conductivity of nanotubes and other carbon allotropes is presented. We show that isolated nanotubes are at least as good heat conductors as high-purity diamond. Our comparison with graphitic carbon shows that interlayer coupling reduces thermal conductivity of graphite within the basal plane by one order of magnitude with respect to the nanotube value, which lies close to that for a hypothetical isolated graphite monolayer.

In Figs. 8.9–8.11 we present the results of our nonequilibrium molecular dynamics simulations for the thermal conductance of an isolated $(10, 10)$ nanotube aligned along the z-axis. In our calculation, we consider 400 atoms per unit cell and use periodic boundary conditions. Our results for the time dependence of the heat current for the particular value $F_e = 0.2\,\text{Å}^{-1}$, shown

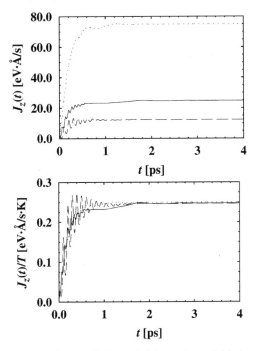

Fig. 8.9. (a) Time dependence of the axial heat flux $J_z(t)$ in a $(10, 10)$ carbon nanotube. Results of nonequilibrium molecular dynamics simulation at a fixed applied thermal force $F_e = 0.2\,\text{Å}^{-1}$, are shown at temperatures $T = 50\,\text{K}$ (dashed line), $100\,\text{K}$ (solid line), and $300\,\text{K}$ (dotted line). (b) Time dependence of $J_z(t)/T$, a key quantity for the calculation of the thermal conductivity, for $F_e = 0.2\,\text{Å}^{-1}$ and the same temperature values. (Reproduced from [33])

in Fig. 8.9(a), suggest that $J_z(t)$ converges within the first few picoseconds to its limiting value for $t \to \infty$ in the temperature range below $400\,\text{K}$. The same is true for the quantity $J_z(t)/T$, shown in Fig. 8.9(b), the average of which is proportional to the thermal conductivity κ according to Eq. (8.29). Each molecular dynamics simulation run consists of 50,000 time steps of $\Delta t = 5.0 \times 10^{-16}\,\text{s}$, or a total time length of 25 ps to represent the long-time behavior.

To study the F_e dependence of the thermal conductivity, we define a quantity by

$$\tilde{\kappa} \equiv \lim_{t \to \infty} \frac{\langle J_z(\mathbf{F}_e, t)\rangle}{F_e T V}. \tag{8.30}$$

In Fig. 8.10 we show the dependence of $\tilde{\kappa}$ on the fictitious thermal force. We have found that direct calculations of $\tilde{\kappa}$ for very small thermal forces carry a substantial error, as they require a division of two very small numbers in Eq. (8.30). Our calculations of the thermal conductivity at each temperature are based on 16 simulation runs, with F_e values ranging from 0.4–$0.05\,\text{Å}^{-1}$.

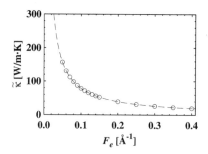

Fig. 8.10. Dependence of the heat transport on the applied heat force F_e in the simulations for $T = 100$ K. The dashed line represents an analytical expression that is used to determine the thermal conductivity κ by extrapolating the simulation data points $\tilde{\kappa}$ for $F_e \to 0$. (Reproduced from [33])

As shown in Fig. 8.10, data for $\tilde{\kappa}$ can be extrapolated analytically for $\mathbf{F}_e \to 0$ to yield the thermal conductivity κ, shown in Fig. 8.11.

Figure 8.11 also shows the temperature dependence of the thermal conductivity of an isolated (10, 10) carbon nanotube. The temperature dependence reveals the fact that κ is proportional to the heat capacity C and the phonon mean free path l. As we discussed, l is nearly constant at low temperatures, and the temperature dependence of κ follows that of the specific heat. At high temperatures, where the specific heat is constant, κ decreases as the phonon mean free path becomes smaller due to Umklapp phonon-phonon scattering processes. Our calculations suggest that at $T = 100$ K, carbon nanotubes show an unusually high thermal-conductivity value of 37,000 W/m·K. This value lies very close to the highest value observed in any solid, $\kappa = 41,000$ W/m·K, which has been reported [5] for a 99.9% pure ^{12}C crystal at 104 K. In spite of the decrease of κ above 100 K, the room-temperature value of 6,600 W/m·K is still very high, exceeding the reported thermal-conductivity value of 3,320 W/m·K for nearly isotopically pure diamond [34]. Another theoretical study has shown that the thermal

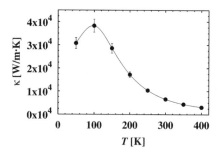

Fig. 8.11. Temperature dependence of the thermal conductivity κ for a (10, 10) carbon nanotube for temperatures below 400 K. (Reproduced from [33])

conductivity of a $(10, 10)$ carbon nanotube approaches $\sim 2,980\,\mathrm{W/m \cdot K}$ along the tube axis [23].

Our theoretical prediction has been confirmed by an experimental measurement of the thermal conductivity of a *single* MWNT at mesoscopic scale [16]. The measured value has been reported to be $\kappa \gtrsim 3,000\,\mathrm{W/m \cdot K}$ near room temperature. We describe some details on this measurement in the next section.

It is useful to compare the thermal conductivity of a $(10, 10)$ nanotube to that of an isolated graphene monolayer and bulk graphite. For the graphene monolayer, we unrolled the 400-atom large unit cell of the $(10, 10)$ nanotube into a plane. The periodically repeated unit cell used in the bulk graphite calculation contained 720 atoms, arranged in three layers. The results of our calculations, presented in Fig. 8.12, suggest that an isolated nanotube shows very similar thermal-transport behavior to a hypothetical isolated graphene monolayer. Whereas even larger thermal conductivity should be expected for a monolayer than for a nanotube, we must consider that unlike the nanotube, a graphene monolayer is not self-supporting in a vacuum. For all carbon allotropes considered here, we also find that the thermal conductivity decreases with increasing temperature in the range depicted in Fig. 8.12.

Very interesting is the fact that once graphene layers are stacked in graphite, the interlayer interactions quench the thermal conductivity of this system by nearly one order of magnitude. For the latter case of crystalline graphite, our calculated thermal-conductivity values are in general agreement with available experimental data [35], [36], [37] measured in the basal plane of highest-purity synthetic graphite, which are also reproduced in the figure. We would like to note that experimental data suggest that the thermal conductivity in the basal plane of graphite peaks near 100 K, similar to our nanotube results.

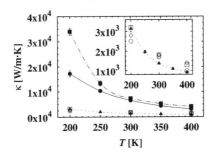

Fig. 8.12. Thermal conductivity κ for a $(10, 10)$ carbon nanotube (solid line) in comparison to a constrained graphite monolayer (dash-dotted line) and the basal plane of *AA* graphite (dotted line) at temperatures between 200 K and 400 K. The inset reproduces the graphite data on an expanded scale. The calculated values are compared to the experimental data of [35] (open circles), [36] (open diamonds), and [37] (open squares) for graphite. (Reproduced from [33])

Based on the described difference in the conductivity between a graphene monolayer and graphite, we should expect a similar reduction of the thermal conductivity when a nanotube is brought into contact with other systems. This should occur when nanotubes form a bundle or rope or interact with other nanotubes in the "nanotube mat" of "bucky-paper" and could be verified experimentally. Consistent with our conjecture is the low value of $\kappa \approx 0.7 \, \mathrm{W/m \cdot K}$ reported for the bulk nanotube mat at room temperature [15], [38].

In summary, we combined results of equilibrium and nonequilibrium molecular dynamics simulations with accurate carbon potentials to determine the thermal conductivity κ of carbon nanotubes and its dependence on temperature. Our results suggest an unusually high value $\kappa \approx 6,600 \, \mathrm{W/m \cdot K}$ for an isolated (10,10) nanotube at room temperature, comparable to the thermal conductivity of a hypothetical isolated graphite monolayer or graphene. We believe that these high values of κ are associated with the large phonon mean free paths in these systems. Our numerical data indicate that in the presence of interlayer coupling in graphite and related systems, the thermal conductivity is reduced significantly to fall into the experimentally observed value range.

8.3 Experiments of Thermal Conduction in Carbon Nanotubes

In the previous section, we discussed theoretical prediction of unusually high thermal conductivity in nanotubes. In this section we discuss the experimental results of thermal conductivity in these materials. Experimentally, carbon-based materials, such as diamond and graphite (in-plane), have exhibited the highest measured thermal conductivity among the known materials at moderate temperatures [39]. The measured value of thermal conductivity of high-quality, 99.9% isotope-free diamond has been recorded up to 40,000 W/mK at 77 K and ~3,000 at room temperature [5]. The in-plane thermal conductivity of graphite is very high: the values of room-temperature thermal conductivity obtained from single crystals and highly oriented pryolytic graphite (HOPG) were reported above 2000 W/m K [40]. Similar to graphite single crystals, a careful study on the axial thermal conductivity of vapor-grown graphite fibers with high-temperature heat treatments (~3000°C) shows that the high thermal conductivity is closely related to the degree of graphitization, that is, the reduction of the grain-boundary density in the samples [41].

The thermal conduction is largely dominated by the phonon contribution in graphite. The electrical conduction is much poorer than that of most metals due to the semimetallic nature of the electron band structure of graphite with greatly reduced carrier density near the Fermi level. On the other hand, the stiff carbon-carbon bonds in graphitized planes increase the speed of sound, and thus, the phonon conduction in the graphite is greatly enhanced.

The thermal conductivity contributed by electrons, κ_{el}, can be estimated experimentally using the Wiedemann-Franz law:

$$\frac{\kappa_{el}}{\sigma T} = L_0, \tag{8.31}$$

where σ is the electrical conductivity and L_0 is the Lorenz number, $L_0 = 2.45 \times 10^{-8}\,(\text{V/K})^2$. In graphite single crystals, it has been found from the thermal- and electrical-conductivity measurements that the phonon contribution of the thermal conductivity, κ_{ph}, dominates κ_{el} and thus $\kappa \approx \kappa_{ph}$ for $T > 20\,\text{K}$ [40].

The discovery of carbon nanotubes [3] has led speculation that this new class of 1D materials could have a thermal conductivity greater than that of graphite due to the long-range crystalline order without boundaries and suppression of phonon-phonon scattering in one dimension [42]. The initially experimental efforts to measure the thermal conductivity of nanotubes have focused on the "bulk" measurements using milimeter-sized mats of nanotubes. The experiments on MWNTs [43] and SWNTs [15] showed the low-dimensional nature of the phonon conduction in these materials. However, the absolute value of the measured thermal conductivity of these bulk samples were two orders of magnitude smaller than theoretically expected values. Later, the improvement in sample preparation was made by aligning nanotubes using a high magnetic field. The measurements of these aligned nanotube samples showed a greatly enhanced measured thermal conductivity in these macroscopic samples allowing the heat flows preferentially along the tubes. Very recently, the group that one of the authors worked with demonstrated mesoscopic thermal conductivity measurements that probed the thermal conductivity of a single isolated MWNT. These results showed that the experimentally measured thermal conductivity of nanotubes is approaching the theoretical predictions as discussed in the previous section [16]. In this section, we discuss briefly the previous bulk thermal-conductivity measurements and then the details of the method and results of the mesoscopic thermal-transport measurements.

8.3.1 Bulk Thermal-Conductivity Measurements of Carbon Nanotubes

In bulk thermal-conductivity measurements of nanotubes, milimeter-sized mat samples are used in a conventional DC measurement setup with differential thermocouples of Ac 3ω method using self-heating of samples. Usually these bulk samples consist of networks of tightly packed nanotube bundles. The filling factor of the sample volume is one of the important factors in the estimation of thermal conductivity from a measured thermal conductance of the samples. Due to the large uncertainty in this filling factor estimation, it is often difficult to obtain the absolute values of the thermal conductivity of the bulk sample accurately. However, the temperature dependence is expected to be less affected by the uncertainty in the filling factor estimation.

The first thermal-conductivity measurements on bulk MWNTs samples were reported by Yi et al. [43] using the MWNTs synthesized by a chemical vapor deposition (CVD) method [44]. The sample consists of the MWNT's diameter ranges 20–40 nm, which correspond to 10–30 graphene walls along the tube axis. A self-heating 3ω method was used in this experiment to measure the thermal diffusivity and specific heat simultaneously. The thermal conductivity was estimated from these quantities considering the filling factor and the density of the sample. Figure 8.13 shows the resulting thermal conductivity of MWNT samples from 4 to 300 K. In the entire temperature range, the measured thermal conductivity, $\kappa(T)$, increases monotonically without any signature of saturation. The room-temperature thermal conductivity is ~25 W/mK, which is much smaller than the theoretically predicted values.

It is interesting to compare $\kappa(T)$ obtained from this bulk MWNT measurement with $\kappa(T)$ observed in graphite fibers. In highly graphitic fibers, $\kappa(T)$ follows a $T^{2.3}$ temperature dependence until $T < 100$ K then begins to decrease with increasing T above ~150 K [41]. The decrease in $\kappa(T)$ above 150 K is due to the onset of phonon-phonon Umklapp scattering. The Umklapp process becomes more effective with increasing temperature as higher-energy phonons are thermally populated. In less graphitic, and thus more disordered, fibers, however, the magnitude of κ is significantly lower and the Umklapp peak in $\kappa(T)$ is absent [41]. This drastic change of $\kappa(T)$ in disordered graphite fiber indicates that the Umklapp phonon scattering is much less important than the grain-boundary phonon scattering in the disordered samples. In the MWNT experiments discussed here, the behavior of $\kappa(T)$ resembles that of disordered graphitic fibers. The room-temperature thermal conductivity is smaller than that of the disordered graphitic fibers, and $\kappa(T)$ does not exhibit a peak due to Umklapp scattering; both properties are consistent with phonon scattering dominantly by the disorders in the samples.

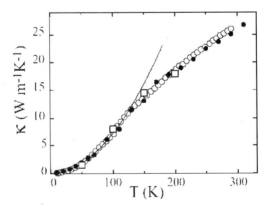

Fig. 8.13. Measured thermal conductivity of MWNTs mat samples at deferent temperatures. Solid line is fit by a T^2 curve up to 120 K. (Reproduced from [43])

Unlike these similarities in the high-temperature regime, the low-temperature behavior of $\kappa(T)$ observed in these bulk measurements is different from that of graphite fibers. At low temperatures ($T < 100$ K), $\kappa(T)$ of the MWNTs increases as $\sim T^2$ (solid line in Fig. 8.13) as opposed to $\sim T^{2.3}$ in graphite. One possible explanation for this difference in low-temperature behavior of $\kappa(T)$ is that the 2D nature of phonon conduction plays an important role in MWNTs. Because the thermal contacts between the MWNTs are only made through the outermost walls (thermal conduction is relatively poor across shells), only the outermost walls of MWNTs can contribute to the thermal-conductivity measurements. Because of the large diameter of MWNTs, phonons in the outermost walls essentially behave like a 2D phonon system. If this interpretation is correct, the magnitude of axial thermal conductivity of MWNTs can be much larger than the measured $\kappa(T)$. We discuss this interesting dimensional cross over in MWNT phonon transport later in this section in connection with the recent mesoscopic measurements.

We now discuss the experimental thermal conductivity of SWNT bulk samples. Compared to the MWNT mat samples, much effort has been made in SWNTs to improve the bulk sample quality for thermal transport measurements. Hone et al. first reported the measured thermal conductivity in an unaligned mat sample that consists of ropes of SWNTs average diameter 1.4 nm [15]. Later, Hone et al. reported much improved results using aligned SWNTs samples [45]. In the latter experiment, the SWNT ropes in the sample were aligned by a suspension deposition in a high magnetic field, followed by annealing the samples at 1200°C to help the tight packing of SWNT ropes in the sample. The very high density samples (about half the crystallographic value) were obtained in this method. A comparative DC technique and an AC self-heating method were used in low-temperature (10–300 K) and high-temperature (300–400 K) ranges, respectively. Figure 8.14 compares of measured thermal conductivity of these aligned and unaligned samples. The thermal conductivity, $\kappa(T)$, was measured along the aligning axis for the aligned sample. Although the temperature dependence of $\kappa(T)$ is similar for both 1.4-nm-diameter SWNT samples, the magnitude of $\kappa(T)$ is very sensitive to the disorderness of samples. In unaligned disordered samples, the room-temperature thermal conductivity is only $\sim$35 W/m K, while the aligned and less disordered sample shows the thermal conductivity higher than 200 W/m K. This observation implies that the numerous junctions in nanotubes mat are the dominant thermal resistance source for the bulk thermal conductivity measurements. At room temperature, $\kappa(T)$ of the aligned SWNT sample is still an order of magnitude smaller than that of diamond or graphite (in-plane). Above 300 K, $\kappa(T)$ increases slowly and then levels off near 400 K. Graphite and diamond, on the other hand, show a decreasing $\kappa(T)$ with increasing temperature above $\sim$150 K due to phonon-phonon Umklapp scattering, as we discussed earlier.

This absence of the Umklapp scattering peaks in $\kappa(T)$ of SWNTs ropes implies that the dominant phonon-scattering source in a rope of SWNTs is

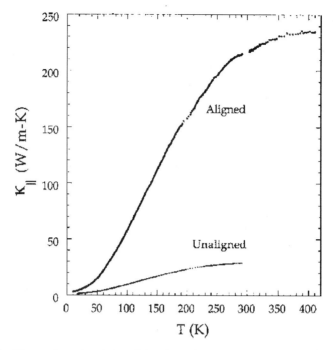

Fig. 8.14. Thermal conductivity of the aligned and unaligned SWNT bulk samples. The thermal conductivity was measured parallel to tube direction for the aligned sample. (Reproduced from [45])

defects or boundary-scattering-related mechanism rather than phonon-phonon interaction. Although an isolated SWNT does not have a boundary that a phonon can scatter off, a rope of SWNT has its boundary that a long wavelength phonon can scatter off. In addition, the restricted geometry of the SWNTs may also affect the Umklapp scattering process. In 1D systems, Umklapp scattering is expected to be suppressed due to the low availability of appropriate phonons for conservation of energy and wave vector [46]. This inherent suppression of phonon-phonon scattering in one dimension can be the alternative explanation of the observed absence of Umklapp scattering peak in $\kappa(T)$. More quantitative work should elucidate this important issue, extending this measurement to higher temperatures and using SWNT bulk samples with different diameters and disorderness.

The low-temperature behavior of $\kappa(T)$ of SWNT samples deserves more attention. As highlighted in Fig. 8.15, $\kappa(T)$ of SWNT bulk samples shows a linear temperature dependence below ∼35 K. This linear T dependence of $\kappa(T)$ at low temperature reflects the 1D phonon band structure of individual SWNTs [47]. At low temperatures, only the four acoustic phonon modes are thermally populated, while at slightly higher temperatures the lowest zone-folded phonon subband begins to be populated [11]. $\kappa(T)$ can be modeled

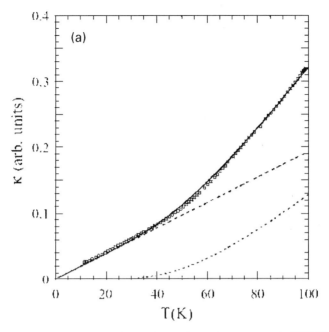

Fig. 8.15. Temperature-dependent thermal conductivity of SWNTs at low-temperature regime. The linear broken line represents the contribution from the acoustic band, and the quadratic dotted line corresponds to the contribution from the first subband in the two- band model (see text for details). The solid line is the sum of the two contributions compared to the measurements (circles). (Reproduced from [47])

using a simplified two-band model, considering only an averaged single acoustic band and one zone-folded subband. In a simple zone-folding picture, the acoustic band has a dispersion relation $\omega = vk$ and the first subband has dispersion relation $\omega^2 = v^2k^2 + \omega_0^2$, where $w_0 = v/R$, and R and v are the radius and sound velocity of the sample, respectively. The thermal conductivity from each band can then be estimated using these dispersion relations [47]. The solid line in Fig. 8.15 shows a fit with this two-band model with the parameters v and $\hbar\omega_0/k_B$ chosen to be 20 km/s and 35 K, respectively [47]. These choices are within a reasonable range considering that the sound speed of two acoustic modes in nanotubes are 15 km/s (twist mode) and 24 km/s (longitudinal mode) [11]. Note that in this model, the contribution from the lowest optical subband freezes out at temperatures below ∼35 K as indicated by the dotted line in the figure. Below this temperature, $\kappa(T)$ shows a linear T dependence. A similar linear T dependence of $\kappa(T)$ was observed in micromachined silicon nitride membrane at very low temperatures (<0.6 K) and has been interpreted as the signature of the quantization of phonon thermal conductance in a 1D system [48]. Although the absolute value of $\kappa(T)$ of an individual SWNT was not obtained in the bulk measurements, qualitatively, this linear T behavior

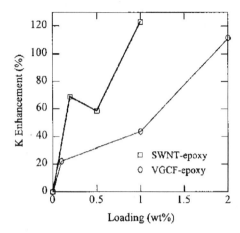

Fig. 8.16. Enhancement in thermal conductivity relative to pristine epoxy as a function of SWNT and vapor-grown carbon fiber (VGCF) loading. (Reproduced from [49])

of the SWNTs $\kappa(T)$ below 35 K strongly suggests that the phonon conduction may be quantized in SWNTs. Extremely restricted phonon-transport channels and absence of the boundary scattering make such quantization occur at relatively elevated temperatures in SWNTs. Thermal-conductance measurement on an isolated SWNT should address this interesting problem in the future.

In addition, SWNTs were used to augment the thermal-transport properties of industrial epoxy [49]. Figure 8.16 shows the thermal conductivity enhancement in this SWNT-epoxy composite material. Epoxy matrixes loaded with 1 wt% SWNTs show a 70% increase in thermal conductivity at 40 K, rising to 125% at room temperature (squares). As shown in the same graph for a comparison, the enhancement due to 1 wt% loading of vapor-grown carbon fibers is three times smaller (circles). These results suggest that the thermal-conduction properties of epoxy composites are significantly enhanced by optimally introduced highly thermal-conductive nanotubes into the epoxy matrix, and SWNTs are much more effective for this than larger-diameter carbon fibers.

8.3.2 Experimental Method for the Mesoscopic Thermal Transport Measurement

Although the previously mentioned bulk studies have provided a qualitative understanding of the thermal properties of these materials, there are significant disadvantages to these macroscopic measurements for understanding intrinsic thermal properties of a single nanotube. One problem is that these measurements yield an ensemble average over the different tubes in a sample. Moreover, in thermal-conductivity measurements, it is difficult to extract absolute values for these quantities due to the presence of numerous tube-tube junctions. These junctions are in fact the dominant barriers for thermal transport in a mat of nanotubes. Most importantly, it is only at the mesoscopic scales where one can study the quantum limit of energy (thermal) transport. In this regard, the mesoscopic thermal-transport measurements are necessary to elucidate the intrinsic thermal properties of nanotubes. Thanks to the advances of modern semiconductor process technology, such mesoscopic experiments on semiconducting devices have been recently demonstrated [48]. Inspired by this work and using novel hybridized synthesis techniques in combination with semiconductor device fabrication techniques, one of the authors recently demonstrated mesoscopic thermal transport in carbon nanotubes [16]. In this subsection, we discuss this novel method, which allows us to probe the thermal-transport properties of nanotubes at mesoscopic scales.

For small-signal thermal-transport measurements, AC thermal transient techniques, such as the self-heated 3ω method [43], [50] are often used to enhance measurement precision. In these methods, an AC heating current with a frequency ω is applied to generate an oscillating thermal energy flow at a frequency 2ω in the sample. Because the propagation of the heat wave is related to the thermal diffusivity and the thermal conductivity of the sample, measuring this wave propagation as a function of ω will provide the values of these quantities [43]. The measurement of the heat wave propagation can be achieved by probing the amplitude and phase of the third-order harmonics (3ω) in the heater voltage. This 3ω voltage fluctuation is caused by the resistance change of the heater itself due to the 2ω fluctuation in temperature. It has been successfully demonstrated that this AC 3ω method can be used to measure the thermal conductivity of bulk nanotube samples [43], [45]. However, it is difficult to apply this 3ω method directly to a mesoscale measurement due to the limited dimension of samples. Especially, for individual nanotubes with a usual length of 1–10 μm, the required frequency to observe the changes in amplitude and phase of heat-wave propagation is more than 100 MHz. Such high frequency thermal-transport measurements are very difficult to realize in a mesoscopic scale. Therefore, a conventional steady-state measurement technique is suitable to probe the thermal conduction through a nanotube at mesoscopic scales.

Figure 8.17 shows a schematic for the steady-state thermal- transport measurements. A sample is clamped between the two thermal reservoirs at the temperatures T_h and T_c, respectively. Thermal energy is supplied from the heat reservoir on the left side so that $T_h \geq T_c$. To measure the heat flow, Q, through the sample, a calibration reference with a known thermal conductance, K_0, is connected between the cold reservoir (T_c) and the heat sink at temperature T_0 ($T_c > T_0$). The thermal conductance of the sample, K_s, is obtained by $K_s = K_0(T_h - T_c)/(T_c - T_0)$. This simple experimental scheme works only if: (1) all the heat flow through the sample flows through the calibration reference, (2) K_s is not too small compared to K_0. The latter condition ensures that the measurement has a high enough signal-to-noise ratio.

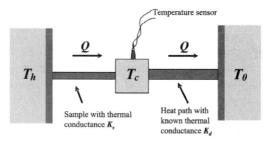

Fig. 8.17. Schematic for experimental setup of steady-state thermal-conductivity measurements.

These two conditions put the following stringent requirements for the thermal conduction measurements at mesoscopic scales. First, the device for measuring temperatures and applying heat should be suspended and free from substrate contact except through small but well-controlled thermal pathways. Therefore, the experiment should be carried out under high vacuum to suppress the residual gas conduction. The radiational heat loss at high temperature ranges should also be considered carefully. These requirements reduce the parasite heat loss from the thermal paths and allow the first condition to hold at mesoscopic scales. For example, the thermal conductance due to residual gas conduction of air, K_{gas}, is estimated from a simple kinetic theory:

$$K_{gas} \approx 10^{-5} \times \frac{P(\text{torr})\, l_c(\mu\text{m})}{\sqrt{T/300K}} \text{ Watt/K}, \qquad (8.32)$$

where P is pressure and l_c is the characteristic length scale of the measuring device. For isolated nanotubes, we expect $K_s \lesssim 10^{-7}$ Watt/K, while $l_c \sim 10\,\mu\text{m}$. Thus $P \ll 10^{-4}$ torr is required for room-temperature measurements. For the thermal-transport measurements in the ultimate quantum limit, $K_s \sim 10^{-12}$ Watt/K at 1 K [48], and hence $P \sim 10^{-9}$ torr is required. On the other hand, the radiational heat loss, K_{rad}, from the surface of the mesoscopic sample is not significant compared to K_{gas}. Due to the greatly reduced surface area in this mesoscopic measurement, $K_{rad} \leq 10^{-13}$ Watt/K even at 300 K for an isolated nanotube sample.

In addition, very thin and long bridges are required to suspend the heaters, thermometers, and samples to achieve the condition $K_s \sim K_0$. For example, a patterned silicon nitride strip with cross-sectional area $1 \times 1\,\mu\text{m}^2$ and length 200 μm provides $K_0 \sim 10^{-7}$ Watt/K at room temperature, which is compatible for an isolated MWNT measurement ($K_s \sim 10^{-8}$ Watt/K), but still too large for a single SWNT measurement ($K_s \sim 10^{-10}$ Watt/K). Furthermore, the metal lines that connect the heater lines and the thermometer should be designed to meet the second condition, especially for low temperatures where electron thermal conduction becomes important compared to the phonon counterpart. For extremely low-level signal measurements, such as the quantized thermal-conductance experiment ($K_0 \sim 10^{-13}$ Watt/K), superconducting leads that completely remove the electronic thermal conduction should be used [48].

Thanks to the advances MEMS technology, the aforementioned suspended devices for the mesoscopic thermal-transport measurement are possible. Recent work on the thermal-conductivity measurement in an isolated MWNT used this advantage [16]. In this experiment, suspended structures were fabricated on a silicon nitride/silicon oxide/silicon multilayer substrate. A low-stress silicon nitride 0.5-μm-thick layer and a 10-μm-thick silicon oxide layer were grown on a silicon wafer by the CVD method. The microscopic Pt/Cr heaters, thermometers, and lead lines were fabricated by electron beam lithography. After fabricating metallic structures, the silicon nitride layer was patterned by photolithography followed by a reactive ion etching that anisotropically etches away exposed silicon nitride layer. In the final step, the silicon oxide sacrificial layer is etched away by HF wet etching, followed by a critical point-drying process. The resulting microdevices were suspended 10 μm above the underlying silicon substrate.

Figure 8.18 shows a representative device, including two 10-μm × 10-μm adjacent silicon nitride membrane (0.5-μm-thick) islands suspended with 200-μm-long silicon nitride beams. On each island, a Pt/Cr thin film resistor,

Fig. 8.18. A large-scale scanning electron microscopy (SEM) image of a micro-fabricated suspended device. Two independent islands are suspended by three sets of 250-μm-long silicon-nitride legs with Pt/Cr lines that connect the microthermometer on the islands to the bonding pads. The scale bar represents 100 μm. The inset shows an enlarged central part of the suspended islands with the micro resistors. The scale bar represents 1 μm. (Reproduced from [16])

fabricated by electron beam lithography, serves as a heater to increase the temperature of the suspended island. These resistors are electrically connected to contact pads by the metal lines on the suspending legs. Because the resistance of the Pt/Cr resistor changes with temperature, they also serve as a thermometer to measure the temperature of each island.

Once such suspended devices are fabricated, the mesoscopic thermal transport in nanotubes and nanowires can be probed as described later. The mesoscopic sized samples are placed on the device and form a thermal path between two suspended islands that are otherwise thermally isolated from each other. Figure 8.19 shows a simple schematic for the heat transfer in such a hybrid device. The islands with heater resistor, R_h, and temperature sensor resistor, R_s, are suspended by beams with the total thermal conductance K_d for each island. A bias voltage applied to one of the heater resistors, R_h, creates Joule heat and increases the temperature, T_h, of the heater island from the thermal bath temperature T_0. Under steady state, there is heat transfer to the other island through the sample with the thermal conductance of the connecting sample, K_s, and thus the temperature, T_s, of the resistor R_s also rises. One can use a linear heat-transfer model to extract K_s and K_d, from the relations of the temperature increases as a function applied to heating power, W:

$$T_h = T_0 + \frac{K_d + K_s}{K_d(K_d + 2K_s)}W; \quad T_s = T_0 + \frac{K_s}{K_d(K_d + 2K_s)}W. \qquad (8.33)$$

It is worth noting that in the derivation of this equation we assumed that each suspended island is in thermal equilibrium at temperature T_h and T_s, respectively. This assumption is valid for $K_s < K_d$, which generally holds for most nanoscale material measurements we discuss.

These suspended structures have been used to measure the thermal conductivity and thermoelectric power of nanotubes [16] and nanowires [51]. Mechanical manipulation similar to that used for the fabrication of nanotube scanning probe microscopy tips [52] has been used to place nanotubes and nanowires on the desired part of the device. This approach routinely produces a MEMS-nanotube/nanowire hybrid device that can be used to measure the

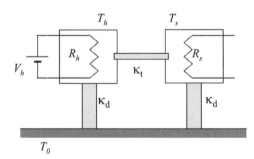

Fig. 8.19. Schematic heat-flow model of the suspended device with a sample across the two islands.

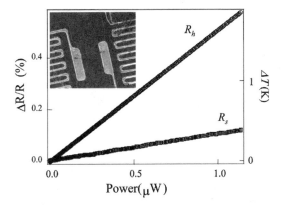

Fig. 8.20. The change of resistance of the heater resistor (R_h) and sensor resistor (R_s) as a function of the applied power to the heater resistor with a silicon nanowire bridging the two islands. The inset shows a detailed SEM image of the device. The scale bar represents $1\,\mu m$.

thermal conductivity and thermoelectric power of the bridging nanomaterials. Shown in the upper panel of Fig. 8.20 is an example of such a device with a silicon nanowire. The silicon nanowire with a diameter ∼50 nm is bridging the two suspended islands with microfabricated heaters (inset). The graph shows the temperature changes of each suspended island connected by this nanowire as a function W at room temperature. From the slopes of R_s and R_h versus W, K_s and K_d at temperature T_0 can be computed using Eq. 8.33. By measuring the samples diameters and length, the thermal conductivity of the sample can be estimated at different temperatures.

8.3.3 Thermal Conductivity of Multiwalled Nanotubes

In this subsection, we discuss the experimental results of mesoscopic thermal conductance measurement of MWNTs using the suspended devices described in the previous subsection. The thermal conductivities of small bundles of MWNTs and an isolated MWNT sample are presented.

Figure 8.21 displays the image of the MEMS-nanotube hybrid device that was used for the thermal conductance measurement of a single MWNT. The MWNT in this device has a 14-nm diameter and a 2.5-μm length of the bridging segment. The thermal conductance, K_s, was measured as described in the previous section in the temperature range 8–370 K. Below 8 K, both R_s and R_h become saturated due to the impurity scattering in Pt/Cr resistors and cannot be used for a temperature sensor. To ensure that the measurement remains in the linear response regime, W was limited to make $T_h - T_0 < 1\,K$ during the measurement. The measured K_s increases by several orders of magnitude as the temperature is raised, reaching a maximum of approximately

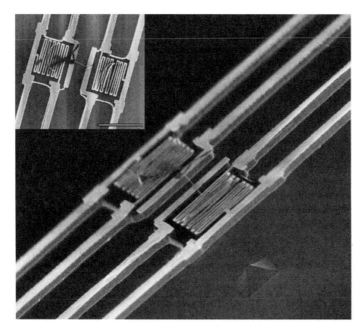

Fig. 8.21. SEM image of the suspended islands with a bridging individual MWNT. The diameter of the MWNT is 14 nm. The inset shows the top view of the device. The scale bar represents 10 μm.

1.6×10^{-7} W/K near room temperature before decreasing again at higher temperatures.

This measured thermal conductance includes the thermal conductance of the junction between the MWNT and the suspended islands in addition to the intrinsic thermal conductance of the MWNT itself. From a study of scanning thermal microscopy on a self-heated MWNT, the thermal conductance of the nanotube-electrode junction at room temperature was estimated; the heat-flow rate from a unit length of the tube to metal electrode at a given unit junction temperature difference was found to be ~0.5 W/m K. Considering the contact length of the MWNT to the electrodes on the islands is ~1 μm, the junction thermal conductance is $\sim5 \times 10^{-7}$ W/K at room temperature. Because the total measured thermal conductance is 1.6×10^{-7} W/K, this suggests that the intrinsic thermal conductance of the tube is the major part of the measured thermal conductance.

To estimate thermal conductivity from the measured thermal conductance, we have to consider the geometric factors of the MWNT and the anisotropic nature of thermal conductivity. The outer walls of the MWNT that make good thermal contacts to a thermal bath contribute more in thermal transport than the inner walls, and the ratio of axial to radial thermal conductivity may influence the conversion of thermal conductance to thermal conductivity. Although the anisotropic electronic transport in MWNTs has been studied recently [54], the anisotropic nature of thermal transport in MWNTs has not been studied to date. Without knowing this anisotropic ratio of MWNT thermal conductivity, it would be the first-order approximation to estimate the averaged thermal conductivity, assuming a solid isotropic material to consider geometric factors. This simplification implies that the thermal conductivity estimated in this work is a lower bound of the intrinsic axial thermal conductivity of an MWNT. Further study to analyze the contribution of individual layers of MWNTs in the thermal transport should elucidate this important issue in the future. Another major factor of uncertainty to determine the thermal conductivity arises from the uncertainty in diameter measurement. A high-resolution SEM was used to determine the diameter of the MWNT in this device. For MWNTs with ~ 10 nm diameter, the resulting uncertainty in thermal conductivity can be as high as 50% of the estimated value.

Shown in Fig. 8.22 is the temperature-dependent thermal conductivity, $\kappa(T)$, of the isolated MWNT in Fig. 8.21. This result shows remarkable differences from the previous "bulk" measurements as described here. First, the room-temperature value of $\kappa(T)$ is greater than 3000 W/m K, whereas the

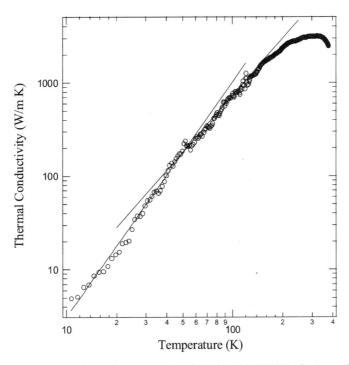

Fig. 8.22. The thermal conductance of an individual MWNT of 14-nm diameter. The solid lines represent linear fits of the data in a logarithmic scale at different temperature ranges. The slopes of the line fits are 2.50 and 2.01, respectively.

previous "bulk" measurement on an MWNT mat using the 3ω method estimated only $20\,\mathrm{W/m\,K}$ [42]. Note that our observed value is also an order of
magnitude higher than that of aligned SWNT samples ($250\,\mathrm{W/m\,K}$) [44] but
comparable to the theoretical expectation, $6000\,\mathrm{W/m\,K}$ as described in the
previous section. This large difference between single-tube and "bulk" measurements suggests that numerous highly resistive thermal junctions between
the tubes largely dominate the thermal transport in mat samples. Second,
$\kappa(T)$ shows interesting temperature-dependent behavior that was absent in
"bulk" measurement. As shown in this log-log plot, at low temperatures,
$8\,\mathrm{K} < T < 50\,\mathrm{K}$, $\kappa(T)$ increases following a power law with exponent 2.50.
In the intermediate temperature range ($50\,\mathrm{K} < T < 150\,\mathrm{K}$), $\kappa(T)$ increases
almost quadratically in T (i.e., $\kappa(T) \sim T^2$). Above this temperature range,
$\kappa(T)$ deviates from quadratic temperature dependence and has a peak at
$320\,\mathrm{K}$. Beyond this peak, $\kappa(T)$ decreases rapidly.

This observed behavior of the thermal conductivity can be understood by
considering the dimensionality changes in the MWNT phonon system and
Umklapp phonon scattering. In a simple model, the phonon thermal conductivity can be written as

$$\kappa = \sum_p C_p v_p l_p, \tag{8.34}$$

where C_p, v_p, and l_p are the specific heat capacity, phonon group velocity,
and mean free path of phonon mode p, respectively. The phonon mean free
path consists of two contributions—$l^{-1} = l_{st}^{-1} + l_{um}^{-1}$, where l_{st} and l_{um} are
static and Umklapp scattering length, respectively. At low temperatures, the
Umklapp scattering freezes out, $l = l_{st}$, and thus $\kappa(T)$ simply follows the temperature dependence of C_ps. For MWNTs, below the Debye temperature of
interlayer phonon mode, $\Theta_\perp$, $\kappa(T)$ has a slight three-dimensional nature, and
$\kappa(T) \sim T^{2.5}$ as observed in graphite single crystals [40]. As $T > \Theta_\perp$, the interlayer phonon modes are fully occupied, and $\kappa(T) \sim T^2$, indicative of the 2D
nature of thermal conduction in an MWNT. From this cross-over behavior of
$\kappa(T)$, we estimate $\Theta_\perp = 50\,\mathrm{K}$. This value is comparable to the value obtained
by a measurement of specific heat of MWNT [43]. At very low temperatures,
one should expect that the phonon transport is quantized as was observed
in [48]. The thermal effects of 1D phonon quantization in a nanotube should
be measurable as $T < T_{1D} = hv/k_B R$, where h is the Plank constant, k_B is
the Boltzman constant, and R is diameter of the nanotube. For an MWNT
with 10-nm diameter, however, T_{1D} is estimated to be $\sim 3\,\mathrm{K}$, which is the
temperature range in which the experiments could not be carried out. Above
this temperature, phonon transport in an MWNT essentially behaves like in
a 2D graphene sheet as we discussed earlier.

As T increases, the strong phonon-phonon Umklapp scattering becomes
more effective as higher energy phonons are thermally populated. Once
$l_{st} > l_{um}$, $\kappa(T)$ decreases as T increase due to rapidly decreasing l_{um}. At
the peak value of $\kappa(T)$, where $l_{st} \sim l_{um}$ ($T = 320\,\mathrm{K}$), we can estimate the

T-independent $l_{st} \sim 500$ nm for the MWNT using Eq. (8.34). Note that this value is an order of magnitude higher than previous estimations from "bulk" measurements [44] and is comparable to the length of the measured MWNT (2.5 µm). Thus, below room temperature where the phonon-phonon Umklapp scattering is minimal, phonons have only a few scattering events between the thermal reservoirs, and the phonon transport is nearly ballistic. This remarkable behavior was not seen in the bulk experiments, possibly due to additional extrinsic phonon-scattering mechanisms such as tube-tube interactions.

It is also interesting to estimate the electrical contribution of thermal conductivity experimentally, κ_{el}, using the Wiedemann–Franz law. The measured electrical resistance of the MWNT is ~ 35 kΩ, and thus, $\kappa_{el}/\kappa \sim 10^{-3}$ at room temperature. This ratio is found to be even smaller at lower temperatures. Therefore, the thermal transport is completely dominant in all temperature ranges in nanotubes.

Finally, we discuss the diameter-dependent effects in $\kappa(T)$ of MWNT bundles. In most materials, decreasing size of the sample increases the surface-to-volume ratio and thus increases the phonon-scattering rate by the surface boundary. This effect is clearly shown in Fig. 8.23(a), which displays measured $\kappa(T)$ of silicon nanowires with different diameters [51]. As the diameter of silicon nanowire decreases, $\kappa(T)$ decreases drastically. This result clearly demonstrates the suppression of phonon transport due to the increased boundary scattering in small-diameter nanowires. The size dependent effect is completely opposite in the nanotube samples. Figure 8.23(b) displays the measured $\kappa(T)$ of MWNT bundles with different diameters. Surprisingly, $\kappa(T)$ increases as the diameter of the bundle decreases, as opposed to the silicon nanowire example. This interesting behavior arises from the peculiar geometry of nanotubes. Unlike nanowires, nanotubes have no boundary at the surface. A seamless graphitized wall of nanotubes provides phonon transport along the

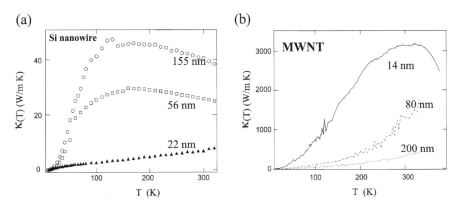

Fig. 8.23. (a) Thermal conductivity of silicon nanowires with different diameters. (Reproduced from [51]); (b) thermal conductivity of MWNT bundles with different diameters. (Reproduced from [16])

tube axis without phonon boundary scattering caused by "surface". Bundling individual nanotubes, on the other hand, creates a new phonon-scattering source by the intertube interaction and reduces the thermal conduction in the sample. Indeed, this experimental observation strongly suggests that the thermal conductivity of an SWNT might be even higher than that of an MWNT due to the absence of intershell phonon scattering as proposed by the theory in the previous section.

8.4 Summary and Future Work

We have presented both theoretical and experimental studies of thermal conduction in carbon nanotubes. The molecular dynamic simulations suggest a thermal conductivity of SWNTs up to 6600 W/mK at room temperature. This unusually high thermal conductivity of nanotubes is associated with the large phonon mean free paths in these systems. The mesoscopoic experimental study in MWNTs indeed confirms this theoretical expectation, showing that the measured values are of the same orders of magnitude. The observed thermal conductivity of an MWNT is more than 3000 W/m K at room temperature and the phonon mean free path is ~500 nm. The temperature dependence of the thermal conductivity shows a peak at 320 K due to the onset of Umklapp phonon scattering. This observation strongly suggest that the phonon Umklapp process might be suppressed in this system due to the effectively reduced dimensionality. Although the theoretically predicted thermal conductivity of SWNTs is in good agreement with the experimentally measured value on MWNTs, we will perform more realistic theoretical calculations on MWNTs to understand some different features on the temperature dependence of thermal conductivity observed between calculations and measurements.

Of particular interest are mesoscopic thermal-transport measurements in SWNTs. The quantization of the phonon degrees of freedom has been shown to modify the heat capacity [45]. This quantization should lead to the thermal-conductance quantization, as shown in other 1D phonon-transport systems at low temperatures [48]. In addition, an extremely long phonon mean free path is expected in an isolated single SWNT due to the absence of intershell phonon scattering and further suppression of Umklapp process in this 1D nanoscale system. Experiments attempting to measure these unique phenomena in SWNTs are underway in one of the authors' group.

Acknowledgments

The authors wish to thank Savas Berber, David Tománek, L. Shi, A. Majumdar, P. L. McEuen and Seung-Hoon Jhi for helpful discussion.

The authors also acknowledge that the work described here has been done in collaboration with Berber, Tománek, Shi, Majumdar, and McEuen. Special thanks go to D. Li and A. Majumdar for sharing the data before publication.

References

[1] M. S. Dresselhaus, G. Dresselhaus, and P. C. Eklund, *Science of Fullerenes and Carbon Nanotubes*. Academic Press, San Diego, 1996.

[2] M. S. Dresselhaus, G. Dresselhaus, and P. Avouris, *Carbon Nanotubes*. Springer-Verlag, Berlin, 1996.

[3] S. Iijima, Helical microtubules of Graphitic Carbon. *Nature (London)* **354**, 56 (1991).

[4] H. W. Kroto, J. R. Heath, S. C. O'Brien, R. F. Curl, and R. E. Smalley, C_{60}: Buckminsterfullerene, *Nature (London)* **318**, 162 (1985).

[5] L. Wei, P. K. Kuo, R. L. Thomas, T. R. Anthony, and W. F. Banholzer, Thermal-Conductivity of Isotopically Modified Single-Crystal Diamond, *Phys. Rev. Lett.* **70**(24), 3764 (1993).

[6] R. Saito, G. Dresselhaus, and M. S. Dresselhaus, *Physical Properties of Carbon Nanotubes* (Imperial College Press, London, (1993)).

[7] J. W. Mintmire, B. I. Dunlap, and C. T. White, Are Fullerene Tubules Metallic? *Phys. Rev. Lett.* **68**, 631 (1992).

[8] N. Hamada, S. Sawada, and A. Oshiyama, New One-Dimensional Conductors: Graphitic Microtubules, *Phys. Rev. Lett.* **68**, 1579 (1992).

[9] R. Saito, G. Dresselhaus, and M. S. Dresselhaus, Topological Defects in Large Fullerenes. *Chem. Phys. Lett.* **195**, 537 (1992).

[10] R. A. Jishi, L. Venkataraman, M. S. Dresselhaus, and G. Dresselhaus, Phonon Modes in Carbon Nanotubules, *Chem. Phys. Lett.* **209**, 77 (1993).

[11] M. S. Dresslehaus and P. C. Eklund, Phonons in Carbon Nanotubes, *Adv. Phys.* **49**(6), 705–814 (2000).

[12] C. Oshima, T. Aizawa, R. Souda, Y. Ishizawa, and Y. Sumiyoshi, Surface Phonon-Dispersion Curves of Graphite (0001) over the Entire Energy Region, *Solid State Comm.* **65**, 1601 (1988).

[13] T. Aizawa, R. Souda, S. Otani, Y. Ishizawa, and C. Oshima, Bond Softening in Monolayer Graphite Formed on Transition-Metal Carbide Surfaces, *Phys. Rev. B* **42**, 11469 (1990).

[14] L. X. Benedict, S. G. Louie, and M. L. Cohen, Heat Capacity of Carbon Nanotubes. *Solid State Comm.* **100**(3), 177 (1996).

[15] J. Hone, M. Whitney, C. Piskoti, and A. Zettl, Thermal Conductivity of Single-Walled Carbon Nanotubes, *Phys. Rev. B* **59**, R2514 (1999).

[16] P. Kim, L. Shi, A. Majumdar, and P. L. McEuen, Thermal Transport Measurements of Individual Multiwalled Nanotubes, *Phys. Rev. Lett.* **87**, 215502 (2001).

[17] C. Uher, Thermal Conductivity of Graphite, In O. Madelung and G. K. White, eds., *Landolt-Börnstein: Numerical Data and Functional Relationships in Science and Technology* **15c** of *New Series, Group III*, 426–448. Springer-Verlag, Berlin (1991).

[18] P. Jund and R. Jullien, Molecular-Dynamics Calculation of the Thermal Conductivity of Vitreous Silica, *Phys. Rev. B* **59**, 13707 (1999).

[19] M. Schoen and C. Hoheisel, The Shear Viscosity of a Lennard-Jones Fluid Calculated by Equilibrium Molecular-Dynamics, *Mol. Phys.* **56**, 653 (1985).

[20] D. Levesque and L. Verlet, Molecular-Dynamics Calculations of Transport-Coefficients, *Mol. Phys.* **61**, 143 (1987).

[21] D. J. Evans and G. P. Morriss, *Statistical Mechanics of Nonequilibrium Liquids*, Theoretical Chemistry Monograph Series. (Academic Press, London, 1990).

[22] D. A. McQuarrie, *Statistical Mechanics.* (Harper and Row, London, 1976).

[23] J. Che, T. Çağın, and W. A. Goddard III, Thermal Conductivity of Carbon Nanotubes, *Nanotech.* **11**, 65 (2000).

[24] A. Maeda and T. Munakata, Lattice Thermal-Conductivity via Homogeneous Nonequilibrium Molecular-Dynamics, *Phys. Rev. E* **52**, 234 (1995).

[25] D. J. Evans, Homogeneous Nemd Algorithm for Thermal-Conductivity: Application of Non-canonical Linear Response Theory, *Phys. Lett. A* **91**, 457 (1982).

[26] D. P. Hansen and D. J. Evans, A Generalized Heat-Flow Algorithm, *Mol. Phys.* **81**, 767 (1994).

[27] D. C. Rapaport, *The Art of Molecular Dynamics Simulation* (Cambridge University Press, Cambridge, 1998).

[28] S. Nosé, A Molecular-Dynamics Method for Simulations in the Canonical Ensemble, *Mol. Phys.* **52**, 255 (1984).

[29] W. G. Hoover, Canonical Dynamics: Equilibrium Phase-Space Distributions, *Phys. Rev. A* **31**, 1695 (1985).

[30] J. Tersoff, Empirical Interatomic Potential for Carbon, with Applications to Amorphous Carbon, *Phys. Rev. Lett.* **61**, 2879 (1988).

[31] J. Tersoff, New Empirical-Approach for the Structure and Energy of Covalent Systems, *Phys. Rev. B* **37**, 6991 (1988).

[32] Y.-K. Kwon, S. Saito, and D. Tománek, Effect of Intertube Coupling on the Electronic Structure of Carbon Nanotube Ropes, *Phys. Rev. B* **58**, R13314 (1998).

[33] S. Berber, Y.-K. Kwon, and D. Tomanek, Unusually High Thermal Conductivity of Carbon Nanotubes, *Phys. Rev. Lett.* **84**, 4613–16 (2000).

[34] T. R. Anthony, W. F. Banholzer, J. F. Fleischer, L. Wei, P. K. Kuo, R. L. Thomas, and R. W. Pryor, Thermal-Diffusivity of Isotopically Enriched $c_1 2$ Diamond, *Phys. Rev. B* **42**, 1104 (1990).

[35] T. Nihira and T. Iwata, Thermal Resistivity Changes in Electron-Irradiated Pyrolytic-Graphite, *Jpn. J. Appl. Phys.* **14**, 1099 (1975).

[36] M. G. Holland, C. A. Klein, and W. D. Straub, Lorenz Number of Graphite at Very Low Temperatures, *J. Phys. Chem. Solids* **27**, 903 (1966).

[37] A. de Combarieu, Thermic Conductivity of Quasi Monocrystalline Graphite and Effects of Irradiation by Neutrons: 1. Measurements, *J. Phys.-Paris* **28**, 951 (1967).

[38] J. Hone, M. Whitney, and A. Zettl, Thermal Conductivity of Single-Walled Carbon Nanotubes, *Synthetic Metals* **103**, 2498 (1999).

[39] G. W. C. Kaye and T. H. Laby, *Tables of Physical and Chemical Constants*, 16th edition (Longman, London, 1995).

[40] B. T. Kelly, *Physics of Graphite* (Applied Science, London, 1981).

[41] J. Heremans, Jr., and C. P. Beets, Thermal Conductivity and Thermopower of Vapor-Grown Graphite Fibers, *Phys. Rev. B* **32**, 1981 (1985).

[42] R. S. Ruoff and D. C. Lorents, Mechanical and Thermal-Properties of Carbon Nanotubes, *Carbon* **33**, 925 (1995).

[43] W. Yi, L. Lu, Z. Dian-Lin, Z. W. Pan, and S. S. Xie, Linear Specific Heat of Carbon Nanotubes, *Phys. Rev. B* **59**, R9015 (1999).

[44] Z. W. Pan, S. S. Xie, B. H. Chang, C. Y. Wang, L. Lu, W. Liu, W. Y. Ahou, W. Z. Li, and L. X. Quan, Very Long Carbon Nanotubes, *Nature (London)* **394**, 631 (1998).

[45] J. Hone, M. C. Llaguno, N. M. Nemes, A. T. Johnson, J. E. Fischer, D. A. Walters, M. J. Casavant, J. Schmidt, and R. E. Smalley, Electrical and Thermal Transport Properties of Magnetically Aligned Single Walt Carbon Nanotube Films, *Appl. Phys. Lett.* **77**, 666 (2000).

[46] R. Peierls, *Quantum Theory of Solids* (Oxford University Press, Oxford, 1955).

[47] J. Hone, Phonons and Thermal Properties of Carbon Nanotubes, In M. S. Dresselhaus, G. Dresselhaus, and P. Avouris, eds., *Carbon Nanotubes* (Springer-Verlag, Berlin, 2001).

[48] K. Schwab, E. A. Henriksen, J. M. Worlock, and M. L. Roukes, Measurement of the Quantum of Thermal Conductance, *Nature (London)* **404**, 974 (2000).

[49] M. J. Biercuk, M. C. Llaguno, M. Radosavljevic, J. K. Hyun, A. T. Johnson, and J. E. Fischer, Carbon Nanotube Composites for Thermal Management, *Appl. Phys. Lett.* **80**, 2767 (2002).

[50] D. G. Cahill, Thermal-Conductivity Measurement from 30 to 750 K: The 3 ω Method, *Rev. Sci. Instrum.* **61**, 802 (1990).

[51] D. Li, Y. Wu, P. Kim, L. Shi, P. Yang, and A. Majumdar, Thermal Conductivity of Individual Silicon Nanowires, *Appl. Phys. Lett.* **83**, 2934 (2003).

[52] H. Dai, J. H. Hafner, A. G. Rinzler, D. T. Colbert, and R. E. Smalley, Nanotubes as Nanoprobes in Scanning Probe Microscopy, *Nature (London)* **384**, 147 (1996).

[53] L. Shi, P. Kim, P. McEuen, and A. Majumdar, unpublished, 2002.

[54] P. G. Collins, M. Hersam, M. Arnold, R. Martel, and Ph. Avouris, Current Saturation and Electrical Breakdown in multiwalled Carbon Nanotubes, *Phys. Rev. Lett.* **86**, 3128 (2001).

Index